T0276256

Praise for *The Rege Guide to Garden Amenuments*

"In a world seemingly focused on the newest miracle from the chemist's laboratory, it is refreshing to find a book that unveils the power of nature in gardening and equips savvy growers with proven tools and formulas to move their practickkes to a new level. Nigel Palmer makes the art and science of life-enhancing amendments accessible to gardeners everywhere."

— **Fred Walters**, *Acres U.S.A.* magazine

"Nigel Palmer's new book is filled with fermentation methods for the garden. It's an exciting new DIY resource for soil regeneration and plant health."

— **Sandor Ellix Katz**, author of *Wild Fermentation* and *The Art of Fermentation*

"Nigel Palmer has come up with a terrific new 'cookbook' for the regenerative (and any other) gardener who desires to build and maintain soil and plant health. The easy-to-follow recipes are designed to satiate any plant's appetite for the appropriate microbiology and nutrients. This guide to 'using locally sourced materials to make mineral and biological extracts and ferments' is to gardening what Julia Child's classic guide is to cooking."

— **Jeff Lowenfels**, author of *Teaming with Fungi* and *Teaming with Nutrients*

"Attention, gardeners! *The Regenerative Grower's Guide to Garden Amendments* will instruct and inspire. Learn to make targeted mineral and biological garden amendments and sprays with this comprehensive and easy-to-understand manual. Nigel Palmer's guide is a classic in the making!"

— **Sally Fallon Morell**, president, Weston A. Price Foundation

"Herbal brews and probiotics keep us in good stead . . . so why not the plants we grow for food and medicine as well? Nigel Palmer offers an enthusiastic exploration of fermented plant extracts and organism cultures to keep our gardens and orchards productive and healthy. Indigenous wisdom presented in a scientific light rocks! Making one's own garden amendments not only saves money but is truly the most effective way to grow. Phytonutrients prime plant immunity and friendly microbes protect from disease. Heed this advice and become a savvier grower today."

— **Michael Phillips**, author of *Mycorrhizal Planet* and *The Holistic Orchard*

"Easy, affordable, and garden-tested amendment recipes to help grow nutrient-dense food? Yes, please! *The Regenerative Grower's Guide to Garden Amendments* made me wonder why I hadn't started making my own amendments years ago. I was barely past chapter 2 before I started collecting egg and oyster shells to submerge in vinegar; comfrey and nettles were already fermenting by the end of chapter 3. Every gardener needs this knowledge!"

— **Chris Smith**, author of *The Whole Okra*;
director, The Utopian Seed Project

"*The Regenerative Grower's Guide to Garden Amendments* is a wonderful book, giving the reader new insight into regenerative horticulture at home— preserving the soil, enhancing biodiversity, and producing healthy food."

— **Jake Fiennes**, general manager of conservation, Holkham Estate

"In *The Regenerative Grower's Guide to Garden Amendments*, Nigel Palmer offers a careful, thorough approach to creating abundance and health in and from the garden. The specifics of innovative and useful techniques, as well as the supporting concepts, are well presented for diligent gardeners to utilize. Nigel's new garden model does a great service in moving forward our ability to feed ourselves in a cooperative effort with nature."

— **Bryan O'Hara**, author of *No-Till Intensive Vegetable Culture*

"This book empowers readers to observe nature's processes at work in the food garden, and to become aware of natural resources, such as weeds and leaf litter, that they can use to make their own mineral and biological amendments. Brilliant! Beautifully crafted by a first-rate teacher able to convey his in-depth understanding, passion, and knowledge, *The Regenerative Grower's Guide to Garden Amendments* will inspire many growers to embark on a never-ending journey of discovery; to grow nutrient-dense food by creating an abundance of healthy ecosystems in the garden."

— **Matthew Adams**, cofounder, Growing Real Food for Nutrition;
former director, Good Gardeners Association

The Regenerative Grower's Guide to

Guide to

Garden Amendments

The Regenerative Grower's

Guide to

Garden Amendments

Using Locally Sourced Materials to
Make Mineral and Biological
Extracts and Ferments

Nigel Palmer

FOREWORD BY John Kempf

Chelsea Green Publishing
White River Junction, Vermont
London, UK

Project Manager: Alexander Bullett
Editor: Fern Marshall Bradley
Copy Editor: Laura Jorstad
Proofreader: Eliani Torres
Indexer: Joel Kaemmerlen
Designer: Melissa Jacobson

Printed in the United States of America.
First printing July 2020.
10 9 8 7 6 5 4   22 23 24

Our Commitment to Green Publishing
Chelsea Green sees publishing as a tool for cultural change and ecological stewardship. We strive to align our book manufacturing practices with our editorial mission and to reduce the impact of our business enterprise in the environment. We print our books and catalogs on chlorine-free recycled paper, using vegetable-based inks whenever possible. This book may cost slightly more because it was printed on paper from responsibly managed forests, and we hope you'll agree that it's worth it. *The Regenerative Grower's Guide to Garden Amendments* was printed on paper supplied by Versa Press that is certified by the Forest Stewardship Council.

Library of Congress Cataloging-in-Publication Data
Names: Palmer, Nigel, 1958- author.
Title: The regenerative grower's guide to garden amendments : using locally sourced
   materials to make mineral and biological extracts and ferments / Nigel Palmer.
Description: White River Junction, VT : Chelsea Green Publishing, 2020.
   | Includes bibliographical references and index.
Identifiers: LCCN 2020019760 | ISBN 9781603589888 (paperback)
Subjects: LCSH: Gardening—Handbooks, manuals, etc. | Plants—Nutrition
   —Handbooks, manuals, etc. | Soil amendments—Handbooks, manuals, etc.
Classification: LCC SB450.96 .P35 2020 | DDC 635—dc23
LC record available at https://lccn.loc.gov/2020019760

Chelsea Green Publishing
85 North Main Street, Suite 120
White River Junction, VT 05001

Somerset House
London, UK

www.chelseagreen.com

To Joan, Miles, Cody

Love, Love, Love

CONTENTS

PART II
Making Mineral and Biological Amendments

FOREWORD

I envision a future where gardening, food production, and agriculture are holistically regenerative, where they quickly regenerate plant health, soil health, the health of livestock, and public health for all the people who consume the exceptionally nutritious food that can be produced in our gardens. In this future time we will manage agriculture from an ecological and ecosystems perspective.

Using this book as a guide, you will be able to produce your own microbial and mineral amendments to improve soil health and the health of your plants without purchased products. The ability to produce our own amendments to regenerate the health of our soils and gardens is important and necessary knowledge to develop sustainable gardening systems.

Sustainable gardening and farming are not yet mainstream practices. A system of growing plants cannot be called *sustainable* before a level of plant health is reached where crops are completely resistant to disease and insect pressure. As long as growers are dependent on synthetic inputs, the production system is not sustainable. Once we have used regenerative practices to achieve levels of ecosystem health that would be considered exceptional today, we can, for the first time, have a legitimate conversation about sustainable agriculture and gardening.

We also cannot have a sustainable agriculture as long as growers are dependent on importing mined or synthesized fertilizers and nutrients. A truly regenerative and sustainable agriculture ecosystem needs to develop methods and technology for tapping into the very large and existing nutrient reserves known to exist in many agricultural soils—reserves that we are largely ignoring. At the time of this writing, the known phosphorus reserves will last perhaps 10 to 15 years at the present rate of use. This mined and refined phosphorus immediately becomes complexed when applied to soil, and only a fraction of it is ever absorbed by crops. At the same time, many agricultural soils contain a supply of phosphorus in the top few inches of soil profile sufficient to last several centuries, in addition to the tremendous reserves in the soils' B horizon.

To have a truly sustainable or even a regenerative agriculture, we need to develop the tools and management systems that harness the amazing power of biology to release the large nutrient reserves held in our soil's geological profile. Microbial populations have the capacity to unlock minerals that are complexed in the soil mineral matrix, sequester nitrogen, and provide nutrients to plants in their most bioavailable form.

In our experience the value and importance of soil biology supersede the balance of soil minerals. It is possible to have a soil analysis that reports perfectly balanced soil nutrients and still have very unhealthy plants when soil biology is dysfunctional. The opposite is seldom the case. Even when soils have imbalanced nutrient profiles, they can still produce very healthy crops, as long as soil biology is vigorous and active. Abundant soil microbiology can overcome challenges of imbalanced chemistry, but perfect chemistry cannot overcome dysfunctional biology.

For all these reasons and more, developing our own nutrition and biological supplements to improve soil fertility and plant health is of foundational importance to developing truly sustainable food production ecosystems. In this book Nigel Palmer points us down the pathway of developing our own solutions for specific challenges we may be experiencing with our crops and soils.

This approach will be a significant part of nutrition management in the future. Regenerative agriculture is recognizing the value and importance of providing plant nutrition that has been metabolized by biology. More and more commercial products are being developed all the time that are based on microbial fermentation processes, and the results in the field speak for themselves. You can develop your own that fit your operation better than a purchased product.

You have the opportunity to participate in the next revolution in plant nutrition.

Enjoy reading and fermenting!

—John Kempf, March 2020

ACKNOWLEDGMENTS

Having a garden was always fun. At some point my wife, Joan, and I realized that growing our own food was one of the best things we could do for our health, and that's when gardening became more deliberate. One day Joan asked me to go with her to hear Dan Kittredge, the executive director of the Bionutrient Food Association, speak about growing food. Dan introduced me to some new ideas and several books and invited me to join him for two days at the Eco-Ag U Workshops at Acres USA, where we listened to Advancing Eco Agriculture founder John Kempf and others speak about eco-growing systems. I was launched into the stratosphere with new ideas and concepts for growing high-quality food. It would take me years to find my own voice regarding this most important subject.

The practices of past cultures seemed to be a great source of information about sustainable growing practices. Lasting civilizations depend on a high-quality, sustainable food system. Recognizing, respecting, and learning about soil ecosystems are clearly required for growing high-quality food over the long term. How was this done in the past?

Rotting plants in a bucket of water, extracting minerals using organic apple cider vinegar, and identifying local mineral resources were common themes and philosophies that were quickly and easily invoked, but the book *Natural Farming Agriculture Materials* by Cho Ju-Young provided me a new level of integration and new understanding of the power of soil biology. Youngsang Cho's *JADAM Organic Farming: The Way to Ultra-Low-Cost Agriculture* took these ideas to another level. Why is this information so hard to acquire?

Teaching sustainable, regenerative gardening classes at The Institute of Sustainable Nutrition (TIOSN) provided a forum to consolidate my thoughts and practices. The many resources reviewed, experiments conducted, data collected, and continuing learning of agricultural ideas there have formed the basis of this book. Making this information available to all has been my impetus.

I would like to thank all the people who document and share their ideas so that others may learn from them and form their own. I so appreciate the many long-standing traditional practices, the detailed scientific analysis of many disciplines, the observations of those paying attention, the dreamers, experimenters, and those who nurture us all.

Thanks to all who try their best and share of themselves.

From Gardening for Fun to Gardening for Health

E verybody knows that food harvested from the garden tastes better than store-bought. Why is this? Is it the freshness, because the food made such a short journey in time and space from garden to plate? Is it the effect of ingesting the diverse microbes that reside on the surfaces of the crop's leaves, fruits, and roots? Perhaps it's the complete absence of chemicals on those surfaces. Or maybe the food tastes so good because of the attention and intent that we express while working in the garden?

I have tended a garden since my early twenties, and my wife, Joan, and I have expanded our gardens over the decades. We began growing berries and herbs along with the vegetables, attracting a wider diversity of pollinators as a result. We always processed some of the harvest for storage, too. Canning tomatoes and drying herbs is just good clean fun. Opening up that jar of salsa in the dead of winter provides a respite from the heavy foods we often eat at that time of year. We called it "summer in a jar." We began to grow enough garlic to supply our needs throughout the year, and then potatoes and more. This was no longer a casual garden; this was our new health program. Learning how to improve the quality of these crops using local sources, as indigenous peoples have done for millennia, was the next step.

In looking for ways to expand this health care program to grow old with, I turned to the world around me. Nature manages to produce beautiful ecosystems—savannas, rain forests, hardwood forests, and more—without the assistance of any products made by humans in factories or laboratories. And for millennia, indigenous cultures have used local nutrient-rich materials as amendments to produce high-quality foods. Learning about these techniques became a passion for me. I already knew the value of manure for gardens, and I practiced cover cropping and crop rotation. Joan and I have always built and managed compost piles to break down the manure from our chickens. Rotating crops is easy in a small garden simply by forgetting what was planted where in previous years.

My next phase of garden experimentation began with putting some weeds in a bucket of water and letting them rot. These turned into very, very stinky concoctions, but it somehow seemed to be the right thing to do at the time. I eventually learned that I could strain these mixtures after the pH dropped to about 5.0, and the strained liquid would be shelf-stable and *not* extremely stinky. I then diluted these products and watered my plants with them—and the results were encouraging. I recalled using vinegar to extract minerals from eggshells when in elementary school—those shells fizzing away, releasing calcium and other minerals into the liquid. Surely I could do something similar. By diluting these liquids and watering my garden with them, I felt that I was amending the soil in ways similar to those employed by growers of the distant past.

As I searched for voices that spoke about the old and new ways of growing high-quality foods, I found another level of information that stirred my interest and passion: the topics of soil mineralization and mineral proportions, soil biology as a plant's digestive system, and the use of a refractometer to measure the quality of fruits and vegetables. The ideas that blueberries all have different levels of antioxidants and other secondary metabolites needed for health, and that better-tasting blueberries have higher sucrose levels (an indication of those antioxidants and metabolites), which could be measured—all of this was empowering. I could put a number on plant health. In addition, I began to learn about the degenerative effects of genetically modified foods on the land and human health, and the deleterious effects of glyphosate on the soil and on human health. This gave me further pause about eating foods purchased from the grocery store. All this learning solidified my conviction to grow my own foods and to do so without purchasing and using any of the agricultural chemicals on sale at the garden center or hardware store.

The first step was to take samples of my garden soil for analysis by a testing lab so that I could understand what the existing proportions of minerals were. With mineral deficiencies and excesses so established I could identify amendments to improve my garden soil. I had to find a lab that offered the type of analysis I needed to measure the macrominerals and microminerals in question. And I had to find a source that provided information about the optimal amounts of these minerals needed in healthy soil. Welcome William Albrecht, Carey Reams, and other visionary agronomists of the 1930s through the 1950s. Once I had read and grasped their work, I went about finding free and low-cost sources of these macro- and microminerals in my local area.

I consulted the geological survey maps from the US Geological Survey—the ones that show what is below the ground, not the kind with contour lines. I was able to locate veins of basalt and limestone (which are good sources of macrominerals such as calcium) on the maps and correlate them with the location of local quarries. I visited the quarries in search of rock dust, which was usually available free of charge. I compared my soil test results with the mineral composition of the dusts to determine which ones would be right for my garden. I also looked for naturally occurring silts on the banks of streambeds after spring flooding, and muck from the bottom of a swamp, bog, or pond. I had samples of these materials analyzed before using them to be sure heavy metals such as lead were not present.

While I was busy with my garden experimentation, Joan was formulating her vision for The Institute of Sustainable Nutrition (TIOSN). Joan is a nutritionist who has become increasingly concerned about nutrition education's narrow focus on the quantitative analysis of food without consideration of food quality. She created a one-year, hands-on certification program structured to encompass the science of nutrition and how soil, food, herbs, and lifestyle influence the health of the body. The program includes culinary skills, kitchen medicine, sustainable foraging, and the importance of sustainable regenerative practices when growing food.

As Joan developed the curriculum, I continued searching for information about using local materials to make amendments. I knew there was more out there—people have been amending soils for millennia—I just hadn't found it yet. Then I came across a book called *Natural Farming Agriculture Materials* by Cho Ju-Young. This book's recipes used indigenous materials in a way that supported the kind of local, sustainable, and regenerative approach I had in mind. The idea of fermenting a specific kind of common plant such as dandelion in order to capture the minerals it contains in a form I could then apply to my garden as a foliar spray was fantastic! And the process that Cho described for capturing indigenous microorganisms (IMO)—local biology, available in the backyard for free—and using it to digest minerals and revitalize the soil ecosystem was just sensational! Tincturing herbs like garlic, ginger, cinnamon, licorice, and angelica root for their medicinal properties to facilitate the digestive characteristics of the soil is an eye-opening and sensible idea. Humans have used tinctures of these powerful herbs for centuries to promote good health; why not use them to facilitate the ecology in the soil? This was the information I was searching for: practical methods that would facilitate nature's processes, rather than chemicals bought in a store designed to destroy parts of the ecosystem. I would never look at stinging nettles, dandelions, purslane, chickweed, comfrey, and valerian the same way again. They were valuable sources of the minerals, especially trace minerals, I was looking for. The recipes in the *Natural Farming* book were difficult to interpret, but the underlying ideas embodied the intuitive, sustainable, regenerative agriculture amendment paradigm I was looking for. *Sustainable* here means there is no waste, no transportation or environmental costs, no heavy machinery purchase required, and none of the hidden costs often overlooked when purchasing a product from a store. Closing waste gaps and utilizing local materials, some of which would otherwise end up in a landfill, in conjunction with these recipes to nurture the soil is a sustainable practice. *Regenerative* means that annual garden practices improve the soil mineral content, biological diversity, and energy flow year after year.

I set to work trying to follow the recipe instructions and began making the amendments. I felt instinctively that these amendments and the process for making them would be good for my garden. But my curiosity about the actual mineral content in the amendment products that I was making got the best of me, and so I sought out laboratories that would conduct mineral content analysis. Having the analysis done was expensive, but it was worth the investment—I was creating a catalog of plant amendment mineral profiles. I also discovered an extensive online database compiled by botanist Dr. James Duke that lists the mineral composition of thousands of plants. (There's more about this database in part 2 of the book.) This discovery brought to light the large distribution of minerals available in different plant types. I also learned that plants are mineral accumulators, and that they accumulate minerals in their tissues in proportions different from those of the soil solution around their roots. I find it amazing and liberating that the minerals needed to grow high-quality food are all around, available for free or at low cost, waiting for growers to understand and incorporate them into their gardening or farming practices.

Understanding the role of soil biology has been equally interesting, and one of the first things I came to understand is how little is known about the life in the soil. I also learned that applying biological amendments made using local biology transforms garden soil. Diverse and ubiquitous biology in the soil ecosystem digests minerals and forms a communication system used by plants. The concepts of biodynamic agriculture introduced by Rudolf Steiner now had context. The powerful plants and manures used to make the biodynamic preparations are from local materials; the processes extract minerals, biology, and energy; and the final products are used to nurture the local soil ecosystem and the plants that grow in it.

I found a second book, *JADAM Organic Farming: The Way to Ultra-Low-Cost Agriculture* by Youngsang Cho, that simplified the concepts presented in *Natural Farming* considerably, reducing processing steps and costs to just about nothing. Most empowering was the recognition that the leaf mold in the woods is the quintessential source of local biology and could be used to inoculate the soil and facilitate the decomposition of plant matter. I had come full circle. I was back to putting weeds in a bucket of water, this time adding a handful of leaf mold biology from my backyard that would digest them, releasing minerals and other compounds. The odors just about disappeared as the anaerobic biological processes decomposed the stinky material on the top of the bucket. Another lesson was that the best mineral amendment with which to feed a plant may be the plant itself. Why not put the carrot tops from the carrots harvested in the summer into a bucket of water, add a handful of leaf mold for biological processing, and use the subsequent mineral concentrations to feed my carrots the following year? Certainly carrots represent the quintessential mineral proportions that carrots want.

Discovering these concepts has led to changes in my home beyond the vegetable garden. I quickly recognized that amending only my garden was shortsighted. It is the entire ecosystem that should be the focus: the lawn, fruit trees, everything. I watched the weeds change as the mineral proportions in the soil changed. I saw the blueberries increase in size, grew sweet-tasting carrots that were as big as my head, and ate Brandywine tomatoes that tasted as if someone had seasoned them to perfection. I had been saving my own seed garlic for nearly 15 years, and the bulbs were getting smaller and more prone to disease in the last couple of years. Once I began using homemade mineral and biological amendments, the quality of my garlic turned around; it became robust and firm again without any indication of disease. Observing the potato bugs arrive, applying a foliar spray, and then noting their disappearance by the following day was amazing. The recognition that what I was doing was not only effective but also in tune with the flows of nature continues to thrill me.

Joan asked me if I would teach my gardening practices at The Institute of Sustainable Nutrition. It became clear that these methods were core to the school's philosophy and were exciting and relevant to the students. Six years of developing curricula and teaching at TIOSN have shaped the pages of this book.

At some point I recognized how important these lessons were to any growers interested in growing in a sustainable manner; I needed to make them available to all. It is my hope that this book becomes stained with nutrient-rich liquids and plant pigments, its pages written on and creased from use.

Real Food Matters

BY JOAN PALMER

As a nutritionist, I often wonder how so many of us can make our living by advising people what to eat. When did food get so complicated? People once ate primarily foods available in season, and they preserved the excess for the cold seasons when fresh produce wouldn't be available. With the exception of those lovely imported foods like coffee, tea, cocoa, and salt, we were dependent on our gardens and the growers around us to provide most of our food. As a result of effective marketing, we learned to feel liberated when we didn't need to spend so much time in the garden or kitchen preparing and preserving food but instead were lured into purchasing chemically constructed foodlike products in eye-catching packages for the sake of convenience. These "foods" bear a price beyond what is paid at the cash register, though—a list of unrecognizable chemicals with unknown long-term effects on health. The flavor of these artificial edibles derived from synthesized additives could make shoe leather palatable. *The Dorito Effect* by Mark Schatzker is a thought-provoking explanation of this concept. A lifetime of this style of eating has left our Western culture suffering with a disturbing range of preventable diseases. Our children, whose diets are full of these edible chemical commodities, are paying the price not only by falling short of their genetic potential but also by developing diseases that historically were experienced by much older people. Children and adults alike are suffering from depression and anxiety in ever-growing numbers. There are many contributing factors to this state of poor health, but our food system is at the core of many health issues we face today.

In 1935 there were close to seven million family farms in the United States, averaging about 155 acres each. These farms and countless backyard gardens yielded a stable local source of fresh seasonable food for most people. During World War II backyard "Victory gardens" produced 40 percent of the fruits and vegetables consumed in the United States. After World War II the number of family farms dropped to below five million with the average farm size growing to 242 acres. Losing small local family farms pushed more of the growing to fewer, larger farms. As a result, food had to be moved greater distances to reach the masses. This centralizing of our food system has continued. In 2019 there were about two million farms in the United States, averaging 444 acres. This trend toward ever-larger, centralized farms was incentivized by tax credits for those operations and the use of chemical fertilizers, herbicides, and pesticides, giving an advantage to industrial farms while disadvantaging smaller organic family farms that do not receive the same subsidies.

It is estimated that more than 50 percent of our food is grown on very large centralized farms of nearly 2,000 acres in size. Small farms that once grew a diverse, sustainable array of food to satisfy the needs of the local communities have been displaced by these large monoculture agribusinesses. Many of the owners of these large farms no longer live in these communities but manage the business from afar.

Most of the produce grown on small farms was harvested by hand and delivered quickly to local markets, looking and tasting fresh. The new practice of growing most of the country's food on massive farms in a few locations and shipping it long distances in all directions raised a number of problems. How can massive quantities of food be transported long distances and still arrive looking fresh and undamaged? This systemic shift in production required changes in the nature of the plants being grown for these far-off markets. Science began the process of hybridization for traits in our food that had nothing to do with flavor and nutrition but instead served the goal of suitability for mechanical farming, days of travel, and superficially "tasty" appearance. Breeders developed crop varieties that could be picked before they were ripe and could withstand the rough handling of mechanical harvesting without bruising. These hybridized fruits could be picked before they had developed their peak color, nutrition, and flavor, and then continue the ripening process, to look beautiful upon arrival in the store. This trait selection fits the needs of agribusiness farming practices, but has left generations of Americans with a less-than-optimal understanding of what healthy, nutrient-dense food is, or what it could taste like.

The adaptive wisdom in plants is amazing. Studies show that fruits develop their full nutrient profile, flavor profile and color while still on the plant. We can pick them early, and they will continue to ripen in color, but they will not further develop their nutrients and flavor in the same way. For example, half a tomato's lycopene and other antioxidants develop in the final stages of ripening. These higher-order compounds provide us with health outcomes beyond just nutrients: cancer protection, heart health benefits, and preservation of cells from oxidative stress. Fully ripened fruits sweeten and decrease in acidity, bitterness, and sometimes toxic compounds. This color, nutrient, and flavor adaptation signals to the environment that the fruit is fully developed and lures humans—and lots of wildlife—to consume it. This is one of the ways these sentient beings spread their seeds, and it has the ancillary benefit of providing humans and animals with potent plant medicine.

Most of the people in the United States today have been shopping in grocery stores, convenience stores, or bodegas their entire life. They have no knowledge of where most of their food comes from or what homegrown food tastes like. Within hours of harvest, exposure to oxygen begins to deplete the nutrients and flavor in our fresh food. Spinach harvested from a megafarm in California can take five days to reach a store shelf, where it may sit for several more days before someone buys it, takes it home, and sticks it in the refrigerator. In this amount of time, spinach will have lost up to 90 percent of its vitamin C and up to half its folate. When food is this depleted of nutrients, the texture and flavor also decline. No wonder so many don't enjoy eating vegetables. Flavor is also one of the ways we have of testing the nutrient density of a plant. If a plant is grown in healthy, mineral-rich soil, teeming with microbial life, it will have the capacity to produce secondary metabolites that improve its nutrient, medicinal, and flavor profile. In my experience teaching people about the link between fresh food and health, I have witnessed many times how surprised children are when they see a carrot being pulled from under the ground for the first time. Even adults reaching into the soil and pulling out their first potato squeal with amazement and delight. Offering a taste of produce fresh from the garden can convert the most skeptical vegetable eaters into eager consumers.

Working in the garden not only produces food and medicine but also has myriad of other health benefits. The act of gardening has been used for decades as a form of therapy for the elderly, the imprisoned, veterans, and others. A student at The Institute of Sustainable Nutrition who had struggled for years with an eating disorder used gardening as a way to help her heal her relationship with food, and she is now working to bring this form of therapy to other people trying to cope with similar issues. And we know it is not just the nutrition that comes from the garden that is healing; we now understand the importance of the bacteria in the soil and on plant surfaces for our health and well-being. Children growing up today have had an overexposure to antibiotics as a result of repeated medical prescriptions, the consumption of factory-farm animal products treated with antibiotics, and the pervasive exposure to the herbicide glyphosate, which was patented as an antibiotic and is found in much of our food and water. This overuse of antibiotics along with the modern-day lifestyle of too much stress, nutritional deficiencies, and overexposure to environmental toxins contributes to our depleted microbiome. This compromised microbiome affects our immune system, mood regulation, social behavior, sleep, memory, digestion, and more. It is no wonder so many are dealing with gut and mental health issues. One round of antibiotics can induce depression in lab mice; think what our children are contending with! By eating from a healthy garden, we are inoculating our guts with healthy bacteria, some of which actually activate cells to produce serotonin similarly to antidepressants. Gardening should be part of every health protocol.

The nutrient difference between food grown at home, or on small, local, healthy farms, and industrial food is an essential conversation when it comes to health. It is why I feel strongly that there needs to be a paradigm shift in the way traditional nutrition programs teach about food and how it relates to health. When I was working on my master's degree in human nutrition, my professors never opened up the incredibly important discussions about the soil, soil microbes, and their influences on food. Too many nutrition programs teach the details of average nutrient contents of food but do not talk about the dependency of those numbers on the health of the soil the food was grown in, the way it was harvested, the time spent shipping it to consumers, and then how it was prepared in the kitchen. Our traditional nutrition education leaves out the amazing and abundant weeds that are accumulators of minerals and a rich source of nutrition for both our gardens and ourselves. How can we talk about food as a source of health and not talk about the poisons applied to the food and soil? Poisons that render plants substantially less healthy and our bodies more depleted as we try to detoxify these foreign chemicals. Our modern nutrition conversation fails to take into account the fundamental importance of the health of the ecosystems where food is grown. This conversation needs to shift.

How to Use This Book

The recipes in this book can be made in a kitchen with simple tools. Most are so simple there is no excuse *not* to make them.

Part 1 of the book provides concepts and definitions, top-level explanations of terms and principles that will provide the context for the use of these recipes. These chapters offer an introduction to topics that can (and do) fill entire books, such as soil biology, mineral nutrition of plants, soil dynamics, plant sap flows, cover cropping, and more.

Part 2 presents the recipes for mineral and biological amendments I make and use in my gardens. Each recipe includes detailed step-by-step instructions, helpful photos, and product use directions, dilution ratios, and storage information. Each begins with a list of needed ingredients and ends with a quick-reference summary. Mineral analysis of some of the products made from these recipes is provided in appendix E.

The appendices are a group of informative resources that expand on the information presented in the book, including plant mineral content, plant nutritional deficiency indicators, a summary of benefits for all the recipes, and a glossary.

This book provides a toolbox that anyone can use to improve the health and diversity of a garden/landscape ecosystem below- and aboveground; the genetic potential of the seeds grown, saved, and replanted year after year; and the quality of the food produced. I focus on the use of locally sourced resources that can be gathered at low or no cost. Some of the recipes produce biology, some minerals. Still others produce complex compounds, and some provide all of the above. Many of the amendments are shelf-stable and can be stored for future use. These mineral and biological amendment recipes are sustainable, regenerative, and effective.

Not only are these ideas stimulating and enlightening, but they make intuitive sense, too. The tools needed to grow high-quality foods are all around us. Realizing this unleashes a world of enjoyment, a bonding with nature, and the flow of the universe. Seeing the changes in the soil, the improved quality of the food grown, and the increase in the number of pollinators affirms the simple, powerful ways of nature.

PART I

Nourishment from the Ground Up

CHAPTER 1

A New Garden Model

We all use models to help explain the world around us. For example, when a traffic light turns green, many of us reflexively hit the accelerator and go. We don't check whether any cars are coming from the left or right, because our mental model assumes that cross traffic has stopped. Another example: At day's end, we don't worry when the sun goes down and it gets very cold, because we have a clear certainty that the sun will rise again tomorrow morning, providing light and warmth. Our mental models define our reality.

We tend our gardens and farm fields based on our mental models about how plants grow. An example of a plant cultivation model is the assumption that simply putting a tomato seed into the ground, and occasionally watering it, will eventually produce a tomato plant that bears a good crop of fruits. Some gardening and farming models require that the soil be tilled and all weeds removed before seeds can be planted. Some models call for regular additions of nitrogen fertilizer and lime. Plants are tenacious, they will grow under the most difficult of circumstances, and even if a cultivation model is far from ideal, the plants may grow pretty well. Many a good-tasting tomato has been grown following conventional gardening models, but are these the most nutrient-dense tomatoes possible to grow? Can the nutrient density of the crops grown be measured? The answers to these questions lie ahead.

Rediscoveries of forgotten or discarded ideas about plant growth—such as that truly healthy plants can thwart soilborne and airborne pathogens, as well as insect pests, and provide significantly higher levels of nutritional value—are the inspiration for a new model for tending gardens and farm fields, one that focuses on the entire ecosystem that the plant grows in. And from these ideas arises a new garden model that makes use of a different set of tools and gardening practices.

The old plant model that most gardeners followed for decades envisions a one-way flow of nutrients from the soil into the plant, with no biological interaction involved. In this model the soil isn't given much importance—it's just a matrix for root growth. Gardeners add nutrients to the soil in the form of fertilizer. Plants absorb those nutrients through their roots, and the plants produce a harvest. Gardeners also remove weeds that might rob the crop plants of the costly fertilizers so carefully applied, and they use sprays to kill pests and disease organisms that might attack the plants.

A new model of plant growth is based on a two-way flow of nutrients between soil and plant and an understanding of a soil rich in biological interactions: the soil as the digestive system for plants. It is a model of symbiosis, recognizing that it is not possible to grow truly healthy plants unless all aspects of the model are in place—minerals in the correct proportions and in forms that the plants can access, appropriate water and sunshine, and a highly functioning soil biology with sufficient energy flow.

In this new plant model, a soil that is rich in a form of organic matter called humus is central. Humus-rich soil provides an environment that will support and shelter a large and diverse biological community, and this is hugely important. Humus-rich soil provides many pore spaces to retain water and allow for good airflow. This gas transfer both in and out is essential because the soil is a living, breathing ecosystem. Humus, as well as clay, is capable of holding mineral ions. Increasing the amount of humus further increases the soil's ability to store these important ions in sufficient quantity to support plant needs for the entire growing season.

Plants can grow in nearly any soil condition, including sterile soil, but when the soil is poor, they cannot produce the complex compounds needed for optimal health. Nor will the plants be able to realize their genetic potential and propagate improved seeds year after year. A healthy and diverse soil biology extends the reach for minerals and water much farther than the plants' root zones. The soil biology not only shares these important resources, but also processes them, providing them to plants as fully assembled complex compounds. According

Improving Energy Flow in Soil

The earth is bathed in energy in the form of electromagnetic radiation (EMR) not only from our sun, but from all the suns in the universe. This energy flow is essential to all life on Earth.

During the day the energy from our sun is used by plant leaves to make food in the form of sugars for itself and the soil underground—think photosynthesis.

The flow of electromagnetic energy is complex and often over-looked when considering health, communication, and our interaction with the world around us. We all know that we are bathed in a magnetic field, which encompasses all of Earth from the North Pole to the South Pole. The magnetic needle of a compass will point to the North Pole nearly everywhere on the surface of the planet. Earth's magnetic field is all around us no matter where we are. Physics defines conditions required for a magnetic field to exist. In addition to the magnetic field, there is also an electric field and current. One cannot exist without the other two. The magnitude of the magnetic field is relatively small so the electric field and the current flow are also relatively small, but all three exist and can be facilitated or inhibited.

Plant saps contain charged particles, the ions that flow in the xylem and phloem pathways. The flow of charged particles is the definition of a current. There is an interaction between the plant sap flow and the current associated with Earth's magnetic field. These energy flows also interact with the soil ecology and may be inhibited or facilitated depending on characteristics of the soil. It's been suggested that the paramagnetic or diamagnetic characteristics of the soil affect the flow of energy through the plant and soil ecosystem.

Another new concept is the value of improving energy flow between the plant and soil. Energy flow is important in biological systems, including humans, but a comprehensive model that can be used to improve energy flow in the soil/plant system may be elusive. An explanation of the concepts of paramagnetism and diamagnetism can scratch the surface of this most important interaction.

Magnetism is the tendency of a material to point in the direction of a magnetic field. For example, the needle of a compass is magnetic, pointing north when exposed to Earth's magnetic field. *Paramagnetism* is the degree to which a material will align with a magnetic field. *Diamagnetism* is the degree to which a material will *not* align with the same magnetic field. Paramagnetism and diamagnetism are quantified by measuring a material's degree of attraction or repulsion to a magnetic field, and this measure is called *magnetic susceptibility*. Paramagnetic materials have positive magnetic susceptibility; diamagnetic materials have negative magnetic susceptibility. Why have I defined all these terms? Simply put, paramagnetic soils will facilitate the flow of electrical energy, while diamagnetic soils will hinder it. This relates back to the fact that efficient energy flow is required to grow healthy plants.

Plants themselves are paramagnetic, acting as antennae. The tiny hairs on their leaves, stems, and stalks, in addition to the leaf surface itself, gather energy that flows through the plant into the roots and soil structure. Take a good look at a tomato plant on a sunny day and notice how those fine hairs on the leaves and stems shimmer in the bright light. Also, over time, it's possible to see that these hairs increase in size and density as the health of the plant increases.

Oxygen is extremely paramagnetic. Thus, developing a soil structure that allows air to penetrate as deeply into the soil as possible is an important step to increasing the paramagnetism of the soil. Adding rock powders that have high magnetic susceptibility may also be beneficial. Volcanic rocks, basalts, and granite may have high magnetic susceptibility. To better understand these concepts, refer to the work of Dr. Philip S. Callahan. (See the bibliography at the end of this book.)

to *Marschner's Mineral Nutrition of Higher Plants*, the fixing of nitrogen, the mobilization of phosphorus, the release of organic acids, and the reduction and oxidation of manganese are some of the actions of soil microorganisms that provide nutrients to plants. Because the plants do not have to expend energy making these compounds gifted by the soil biology, they can use that energy to make even higher-order compounds called secondary metabolites, which are needed for optimal health. Biological diversity establishes resilience and stability by the redundancy of organisms to complete tasks within the soil and also keeps single organisms (pathogens) in check.

The two-way flow of nutrients between plant and soil has radical implications. It's long been known that plants take up water and nutrients from the soil through the xylem pathway, a component of a plant's internal nutrient transport system. But plants also transport sugars produced in their leaves both upward and downward through the phloem pathway, another component of this transport system. Plants release as much as 25 percent of the sugars they produce via photosynthesis through their roots into the soil, thus feeding soil biology. In exchange, the biology provides needed nutrients and water to the plant through the xylem pathway. The soil biology is very adept at breaking down soil minerals into forms the plant can use, directly providing many compounds that the plant would otherwise have to produce itself. As the health of the plant increases and the percentage of sugar within the plant sap increases, insects and pathogens that might otherwise prey on the plants are unable to affect it. The plant is no longer a food source for insects, because the insects do not have the enzymes in their digestive system needed to break down these sugars. Pathogens are unable to penetrate the robust waxy cuticle and epidermis of the plant leaf and thus cannot infect it. (This concept is explained in greater detail in the "Foliar Sprays and Drenches" section in chapter 3.)

With this new plant model as a guide, the gardener or farmer may participate with the natural cycles in the garden, working with nature by making and applying amendments that introduce minerals and a local, thriving soil biology. These amendments provide the spectrum of needed minerals in plant-available forms, and increase the paramagnetic nature of the soil to improve the flow of energy between plant and soil. Making these amendments from locally available materials and applying them to nourish the soil and the plants is the main theme of this book.

The Soil/Plant Model

The new garden model encompasses three overlapping models: a soil mineralization model, a soil biology model, and a plant circulation model.

Soil Mineralization Model

Volcanoes are Earth's soil makers. The movement of tectonic plates and subsequent buildup of energy within the Earth's core are relieved when the liquid center breaks through the surface crust and is unleashed into the sky. Material from the internal mantle spews into the air. Pulverized mountaintop rock is jettisoned as well. This debris covers the land with layers of liquids, fine particles, and larger rocks full of constituent minerals needed to build a thriving soil ecosystem.

Further degradation of volcanic products occurs over time by weather, water, and chemicals. These products are transported by the forces of gravity and wind, as well as the flow of water. Cycles of glacial activity, ice ages, and the advance and retreat of continental ice sheets grind and translocate materials. And of course, plants have been bringing up minerals from the soil and distributing them onto the soil surface for hundreds of millions of years. Mineral-rich soil exists all around the world as a result of all this activity.

Seeds flourish in this mineral-rich environment species by species, each establishing new networks of mineral exchange pathways feeding energy into the soil in the form of sugars. Some of these plants are generally referred to as weeds, but I call them specialists in the area of soil building, experts in growing where other plants will not, mineral accumulators that convert the raw minerals of volcanic spew into usable compounds. Living plants inject compounds into the soil structure in the form of root exudates. When these plants die, their spent foliage deposits some of these compounds on the soil surface, and they add still more compounds within the soil as the roots decay. These compounds include soil-enhancing products like humus and chelated nutrients in plant-available form that subsequent plants and the entire soil ecosystem may utilize in the future.

Too much or too little of just about anything is not good, and minerals in the soil are no exception. Scientists have long understood that there are desired optimal proportions of minerals in soils for the cultivation of crops, but this is not often emphasized in the old plant model paradigm. And although there are general guidelines describing optimal mineral proportions, the nature of these percentages are constantly evaluated and discussed. The identity and number of essential mineral elements plants need to thrive are a subject of debate. *Marschner's*

Mineral Nutrition of Higher Plants lists 14 essential minerals. Another reference text, *Mineral Nutrition and Plant Disease*, suggests at least 16. The list grows as more is learned about these complex ecosystems. The relative proportions of macrominerals (those needed in relatively large amounts) and trace minerals (those needed in much smaller amounts) in the soil are important—there are optimal proportions that form target values to work toward when implementing an amendment strategy.

A full explanation of mineral needs and mineral uptake by plants is beyond the scope of this book. Many recognize the work of William A. Albrecht as the standard for soil mineral composition. *The Ideal Soil*

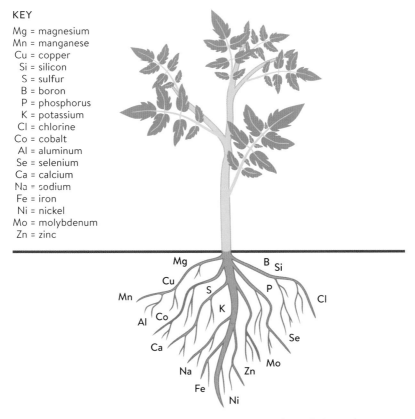

KEY
Mg = magnesium
Mn = manganese
Cu = copper
Si = silicon
S = sulfur
B = boron
P = phosphorus
K = potassium
Cl = chlorine
Co = cobalt
Al = aluminum
Se = selenium
Ca = calcium
Na = sodium
Fe = iron
Ni = nickel
Mo = molybdenum
Zn = zinc

Soil is much more than just dirt. The current state of knowledge indicates that soil must contain the range of mineral elements shown here, and in specific proportions, for optimal plant growth. As our knowledge grows, this list will likely expand.

2014 by Michael Astera offers a useful interpretation of Albrecht's work. *The Intelligent Gardener* by Steve Solomon presents a more simplified explanation. Refer to the bibliography for additional references.

Collecting soil samples and having them analyzed by a soil testing lab is a great place to start learning about the mineral composition of your soil. The test results will provide a measure of existing soil mineral content and, in some cases, the proportions of these minerals within the soil. This information may be used to determine which minerals may be present in excessive amounts and those that are deficient. It's important to understand that mineral excesses can be as serious a problem as mineral deficiencies. (Instructions on how to conduct a soil test are in chapter 4.) Keep in mind that just because minerals are present in the soil does not necessarily mean that plants can access those minerals.

The *exchange capacity* of a soil is a measure of its ability to hold on to and release minerals in ionic forms. As you may remember from high school science, an atom that has gained an electron is called an *anion* and has a negative charge; an atom that has lost an electron is called a *cation* and has a positive charge. Clay and humus are components of soil that have a colloidal structure that is capable of holding on to negative and positive ions. The amount of these colloids in a soil defines that soil's exchange capacity. Exchange capacity is a fascinating subject, and for those interested in a more thorough discussion of the subject I recommend *The Nature and Property of Soils* by Ray Weil and Nyle Brady.

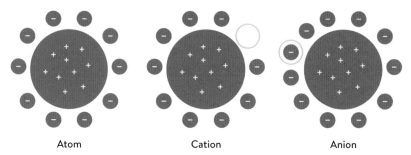

| Atom | Cation | Anion |

An atom has a neutral charge with equal numbers of protons and electrons. A cation has a net positive charge, with more protons than electrons. An anion has a net negative charge, with more electrons than protons.

A high soil exchange capacity enables the storage of mineral ions in forms plants need and use throughout a season. Soils with low exchange capacity may not be able to hold mineral ions throughout the entire growing season. Crops that succumb to disease late in the season are often those that have developed nutrient deficiencies as a result of the soil's inability to hold nutrients or water all the way through. If soil has low exchange capacity, growers can intervene in a beneficial way by applying mineral-rich foliar sprays to provide for the nutritional needs of plants in the short term. (I explain this in more detail in "Mineral Amendment Sources" in chapter 2.) Mulching, composting, and using cover crops to help retain water and to increase the soil's exchange capacity over the long term are also helpful (see chapter 3).

Soil test results can serve as a guide in figuring how much of which minerals need to be added to the soil for optimal plant growth. This involves some mathematical calculations, and those are explained in the "Garden Math" section of chapter 4. There are various application methods for minerals needed in large amounts, such as calcium, sulfur, and magnesium, and other methods for trace minerals like zinc, cobalt, and boron that are needed in very small amounts. Optimizing soil mineral proportions is a long-term activity that may take years.

Soil Biology Model

Gardeners, farmers, and even scientists often exclude the soil ecosystem from their garden and farm models. This may be because there is so much we don't understand about the soil ecosystem. Scientists have not yet identified most of the biological players in the soil, let alone developed an understanding of their functions and interactions.

The model of soil as a digestive system is a good reflection of the current understanding of the symbiotic relationship between plants and the soil ecosystem. The soil ecology not only breaks down minerals into plant-available forms but also manufactures and delivers enzymes and other valuable compounds to plants, all of which contributes to the health of the plants and the ecosystem at large.

Tens of thousands of species of soil fungi have been identified, but at least a million more still await discovery. Plants and some types of fungi form symbiotic relationships in which the fungi receive nutrients in the form of substances exuded by plant roots; in exchange the fungi provide water and nutrients to the plant. Fungi produce a network of threadlike growth called *mycelia*. These long filamentous structures are capable of transporting minerals and water over long distances. When a plant's root system has a relationship with fungal mycelia, it increases the plant's functional capacity to reach through the soil for minerals and water, and this translates to increased nutrition and drought resistance for the plant. This relationship of roots and mycelia is called a *mycorrhiza*, and such fungi are called *mycorrhizal fungi*. Fungal filament strands interconnect the roots of separate plants in the soil ecosystem, thus transferring nutrients from one plant to another! Fungal mycelia also act as communication networks that alert the community of plants in a forest area or a field to the presence of pests or pathogens, and this enables a systematic response by the entire ecosystem. Fungi have the ability to decompose decay-resistant organic components found in soil such as lignins, starch, gums, and cellulose. Fungi also play major roles in the formation of humus. The decomposition of materials by fungi is called *humification*.

An example of the vast amount we do *not* yet know about soil biology are the recently discovered life-forms called archaea. First identified at the end of the 20th century in the hot geysers of Yellowstone Park and the high-pressure, lightless environment of the ocean floor, these organisms were initially thought to be bacteria. However, they turned out to be a distinct life-form. As a result, taxonomists changed the biological classification system, adding archaea as their own domain. (The other two domains are bacteria and eukaryotes—fungi, plants, and animals.) Scientists have determined that archaea are ubiquitous in the soil and contribute to the process of nitrogen fixation, yet rarely are the contributions of archaea even mentioned in agricultural scientific literature.

Bacteria, the oldest form of life on Earth, are also ubiquitous in the soil, with species numbering in the hundreds of thousands or more. Humans have named only about 10 percent of them; many of their functions and interactions are not yet fully understood. One major role of soil bacteria is to decompose organic matter into nutrient forms that plants can utilize. The decomposition of materials by bacteria is called *mineralization*.

In order to have a thriving soil ecosystem, food and housing must be available to support the bacteria, archaea, fungi, nematodes,

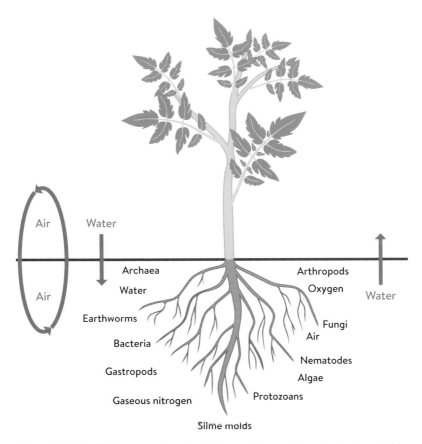

The soil is full of living organisms that help provide nutrients in plant-available forms and digest organic matter to form humus. These living things need a steady supply of water, food, and good air exchange in the soil; atmospheric gases are also essential for their metabolic activity.

protozoans, arthropods, earthworms, and other life-forms that live there. The basic food source of the soil ecosystem is combined carbon. Important sources of combined carbon for garden and agricultural soils are mulches, cover crops, fully decomposed compost, and the root exudates of the plants growing in the soil. These materials also provide housing (shelter) for the soil biology, storage sites for minerals made available by the soil ecology, and pore spaces where water can be stored and through which air can flow. The soil solution consists of the water and water-soluble constituents found within a soil. All these characteristics define a soil's *tilth* and create the environment needed for the soil ecosystem to thrive.

Bacteria, fungi, and archaea are the food sources for larger soil predators such as nematodes and protozoans, which in turn become food sources for even larger predators. The excretions, wastes, and carcasses of this entire ecosystem, from bacteria to beetles, are absorbed into the soil solution, becoming the stuff of good plant nutrition, including minerals in plant-available forms.

A thriving biology even has the capacity to digest rock dusts, bones, shells, and other materials added to the soil to balance mineral proportions. This is why applying both biological and mineral amendments to soil and plants at the same time is a good idea, and one that I'll recommend many times in the course of this book.

The dynamics of the soil ecosystem are fascinating. As an example, consider what happens within the soil when it rains after a dry spell. During dry periods, the soil solution drains down through the spaces within the soil structure to deeper levels, and it also evaporates and leaves the soil as water vapor. However, water confined in the smaller spaces within the soil structure is stored for the future, increasing reserves of water that the plant can utilize during times of drought. The more humus and tilth the soil has, the more such small hiding places for water exist. Also, if the soil surface is well covered by green growth, that slows the rate of evaporation from the soil surface. (That's why bare soil dries out faster than mulched soils or soil where a crop is growing.)

During the dry days after a rain event, there is more air, and thus more oxygen, within the soil, and aerobic biology will prosper and increase their populations. When it rains again, water seeps into the soil. If there's sufficient rain, the soil will become saturated, with all the pore spaces filled by water. This deprives aerobic organisms of the oxygen they need to survive. Huge numbers of organisms may die, and this can actually benefit the soil ecosystem, providing nutrients for plants through the soil solution and also for use by anaerobic forms of biology that will now have an opportunity to thrive and increase their populations. When the glut of water subsides and the air returns, the anaerobes will decline, thus providing a food source again for the plants and the aerobic organisms. And the cycle repeats. There are also *facultative bacteria*, which have the ability to survive in both anaerobic and aerobic environments. Facultative bacteria make ATP by respiration in an aerobic environment and switch to fermentation in an anaerobic environment. Populations of facultative bacteria tend to remain steady during wet and dry conditions.

A practical application of this phenomenon is the use of irrigation to control pathogens on a larger scale. Entire fields may be flooded with water and left saturated for several days in order to eliminate aerobic pathogens that accumulate over time. After the water drains, these fields can be replanted and crops will grow well, with no pathogens to trouble them for some period of time.

In farm fields (and gardens) where conventional practices have been used, the soil is often devoid of biological diversity. One reason for this is tillage, which disrupts soil biology by tearing up the fungal mycelia and releasing carbon into the atmosphere. Leaving soil bare after it has been tilled causes it to dry out, resulting in the further death of soil biology. In addition, insecticides, fungicides, and other chemicals have the unintended effect of killing many beneficial organisms. Learning about nature's diverse soil ecology leads to understanding that the concept of killing parts of the soil ecosystem in order to facilitate the growth and health of the plants grown in this soil ecosystem is fundamentally flawed

The Remediating Power of Biology

The power and utility of soil ecosystems and the plants that grow in them have become recognized as tools that can be used to clean up (remediate) environmental hazards. The decomposing, regenerative effects of the processes called bioremediation, mycoremediation, and photoregeneration have been harnessed to digest oil spills, chemical contamination, plastic wastes, and nuclear waste. These are exciting and hopeful examples of the power of the complex ecological digestive systems present in the soil.

Many farmers, gardeners, and scientists have overlooked the importance of supplying the soil with a diversity of biological amendments, possibly because of our limited understanding of the individual and aggregate functioning of soil bacteria, fungi, and archaea. But they are essential, and the changes that come about in gardens and fields when the biological amendment recipes in part 2 are applied are nothing short of amazing.

Plant Circulation Model

Sap flows through two narrow pathways in plants—the xylem and the phloem. Communication between the two pathways occurs through a thin membrane that allows transfer of nutrients and other compounds. Nearly every cell in a plant sits within a few cells' distance of this membrane and thus can receive the stuff of life. These pathways also form a communication network between the shoot tips and the root tips. Plants send signals about the nutritional status of the plant's *sinks* (sites where energy is used or stored) and regulate soil biological function and the distribution of nutrients based on nutritional needs. The three types of plant sinks are new shoots, fruits, and roots. Nutrients channeled to the roots feed root tip growth and, through root exudates, the soil biology as well.

XYLEM FLOW

The xylem pathway moves the soil solution water and nutrients from the roots up toward the top of the plant. Solute concentrations decrease along the way, depending on plant type, growth phase, and plant nutritional needs. Most of the nutrient-rich solution is transported to the leaves. The movement of solution through the xylem pathway does not require any expenditure of energy by the plant! A force called *osmotic pressure* pushes the soil solution into the root; this force results because there is usually a higher concentration of nutrients inside the root, so the higher water containing soil solution moves into the root, diluting the concentrations. The forces of cohesion and adhesion act to move plant sap upward through the xylem as well. *Cohesion* is the tendency of water molecules to stick together; *adhesion* is the tendency of water molecules to stick to a surface. Another very strong driving force is *transpiration*, which is the flow of water vapor out of leaves into the atmosphere during the day when the sun is bright and the air dry.

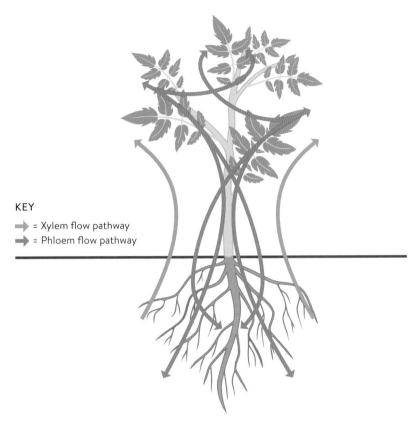

KEY

➡ = Xylem flow pathway
➡ = Phloem flow pathway

Xylem flow carries water and minerals from the roots upward into the plant; phloem flow carries nutrients from the leaves to the plant sinks.

There are other energy-related drivers that are not fully understood—for example, the interaction between plants and the electrical forces associated with Earth's magnetic field.

The amount of xylem flow varies throughout the day as a result of changing temperatures, relative humidity, and sunlight. Minerals that enter the xylem pathway from the soil solution are in ionic forms and include anions of nitrogen (NO_3^-), sulfur (SO_4^{2-}), phosphorus ($H_2PO_4^-$), and chlorine (Cl^-); as well as cations of nitrogen (NH_4^+), calcium (Ca^{2+}), magnesium (Mg^{2+}), potassium (K^+), manganese (Mn^{2+}), sodium (Na^+), and zinc (Zn^{2+}). Boron and silicon are metalloids—they exhibit properties somewhere between those of metals and nonmetals—and they are found mostly in the forms of $B(OH)_3$ and $Si(OH)_4$.

The composition and concentrations of xylem sap depend on plant species, elements in the soil solution, the ability of the roots to assimilate the nutrients in the soil solution, the amount of water in the soil solution, seasonality, and the phase of plant growth. The plant is in charge.

PHLOEM FLOW

The phloem pathway moves nutrients, primarily sucrose and organic compounds, all around the plant; nutrients can move upward or downward in the phloem. Nutrients enter this pathway from the plant stem and leaves and are moved to the major growth sinks of the plant: new shoots, roots, and fruits. Flow is driven by osmotic pressure and the mass flow of nutrients. As the concentrations of water and nutrients change within the different parts of the plant, so does the direction of flow, and thus phloem flow is bidirectional. Nutrients are thus cycled throughout the plant. Phloem nutrient mobility is high for potassium, magnesium, phosphorus, sulfur, nitrogen, sodium, and chlorine; intermediate for iron, zinc, copper, boron, and molybdenum; and low for calcium and manganese.

Inside plant stems, the xylem tissues and phloem tissues are in close proximity, with a permeable layer called the cambium between that allows for two-way communication across the channels. This communication is important, as the xylem flow is mostly directed to sites of highest transpiration and not necessarily to the sites of highest nutritional needs such as fruits or seeds.

The transfer of organic and inorganic compounds between the xylem and phloem pathways is driven by concentration gradients along the full length of these pathways from roots to new tip shoots. The rate of xylem flow influences the transfer rate of these compounds. When the sun is bright during the day and the rate of transpiration is high, the rate of the xylem flow is also high. In the evening, when transpiration is low or nil, the xylem flow is at a low and the energy created during the day's photosynthesis is distributed between the xylem and phloem flows and throughout the plant. This is one reason why foliar sprays are best applied in the evening or early morning; this is explained in more detail in chapter 3.

Learning how minerals flow through the xylem and phloem helps in understanding the best way of supplying nutrients to plants. Two possible

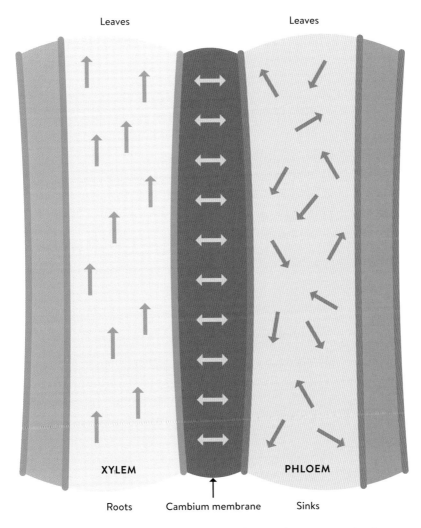

The xylem and phloem channels constitute the circulation system in plants. Nourishment can flow in both directions across the membrane between the two channels, and this communication allows the plant to adjust nutrient concentration needs.

delivery methods are foliar spraying and drenching the soil. Foliar spraying applies amendments as a spray directly to leaves, stems, and bark. Drenching applies amendments as a liquid directly to the soil and root systems via the soil solution. With this in mind, consider the example of calcium. The target soil mineral proportion of calcium is 68 percent. One reason this large proportion of calcium is required in the soil is because plants absorb calcium primarily through the xylem flow path, from the roots up into the plant tops. If the soil mineral proportion of calcium is low, plants will suffer. A regular drench of diluted vinegar extraction of eggshells (which are high in calcium) would be an effective way to provide calcium to plants throughout the growing season. This extracted calcium is in a form that the plant can utilize; the drench liquid mixes with the soil solution, and the calcium is taken up by the plant through the root system into the xylem pathway. Calcium has low mobility in the phloem pathway, but nonetheless a foliar spray may still be effective.

All the amendments described in this book supply a broad spectrum of minerals and other compounds in plant-available forms. If an amendment is applied to the leaves, its content will enter the phloem pathway; when applied to the soil, the contents will move through the xylem pathway.

Putting the Models Together

Plants live symbiotically with soil microorganisms. Plants release 20 to 60 percent of the sugars created during photosynthesis into their roots. Root exudates create an environment around the roots that attracts specific soil microorganisms. Facilitated by these root exudates, the population density of these microbes increases quickly by several orders of magnitude, which increases rates and area of local soil decomposition available to the plant root structure. Each plant type has specific soil biology populations that it supports. As Jeff Lowenfels explains in *Teaming with Microbes*, the content of root exudate changes during the season to manipulate populations of bacteria, archaea, and fungi to suit the plants' changing nutritional needs.

I think these concepts shed a new light on the benefits of companion planting and why some plants do not do well when planted near others. Companion plants stimulate compatible soil biology, making the sum of their outputs more beneficial than either plant could achieve alone. Conversely, plants that promote incompatible soil biology may experience a diminished supply of nutrition and grow poorly together.

In order to produce the healthiest plant, one that reaches its full genetic potential, the plant's root system must interact with a healthy

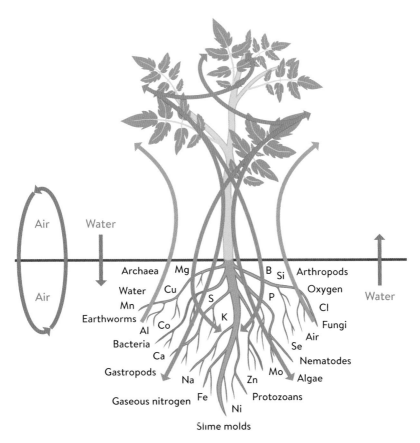

The new garden model includes a spectrum of soil minerals in specific proportions, a thriving soil biology that forms symbiotic relationships with plant roots, and a sufficient flow of water, air, and energy within the soil. Soil biological diversity becomes a measure of ecological function and capacity.

soil ecology and have minerals available in the correct proportions. By selectively feeding the soil biology, plants promote the types of biology that will break down soil minerals into plant-available forms and provide other needed compounds. Thus, we can say that plants have the capacity to create the soil environment in which they can thrive. In this paradigm shift we begin to understand that in nature, acid-loving plants such as blueberries do not magically find acidic soil to grow in. Rather, the soil biology that the blueberry plants support causes a shift to an acidic soil pH! Amazingly, the pH in the rhizosphere (that area of the soil where biology interacts with the roots of plants) may have a pH as much as two units different from the surrounding bulk soil pH and that within a given root system of an individual plant the pH may differ by more than two units along the root axis or between lateral and primary roots.

Scientific Research and Soil Health Variations

With a garden model that so beautifully and naturally meets the needs of growing plants, what led gardeners and farmers to begin using chemical fertilizers and pesticides in the first place? The answer to this question could fill a book on its own, and it might include chapters on politics, the rise of corporate influence in agriculture and horticulture, the power of advertising, and much more. I won't discuss those topics here, but I do want to reflect on the role that scientific research has played in creating misleading information about the efficacy of fertilizers and various forms of pesticides, herbicides, and fungicides.

Validated scientific research requires demonstration of a statistical difference between the results of a control group and an experimental group. The control group defines what is normal, and the experimental group defines the effect of that which is being studied. Showing that there is a statistical difference between these two data sets validates the study and the effect of what was applied to the experimental group. For example, to evaluate the effect of a specific insecticide, a control group of plants would be grown in a specific set of circumstances. The experimental group of plants would be grown in the same set of circumstances, but it would also be treated with the insecticide. Researchers would collect data, such as plant height and plant dry matter weight over time. The data would then be statistically analyzed and compared to determine whether there are differences between the control group and the experimental group. It is assumed that the control group captures all the sources of variation to be considered in the study and that the variation within the control group test data defines the standard deviation of the experiment. The standard deviation of the control is then compared with that of the experimental group to determine whether the differences in the test results are the result of the use of insecticide or are simply due to the sources of variation that could be at play. Without a proper measure of all sources of variation within the data set of the control, comparisons with the experimental data set become questionable.

Sources of variation are the bane of scientific research. When it comes to agricultural research, most studies ignore a significant source of variation that has a potentially large impact on study results—soil health! Variations in soil ecology and the mineral proportions available in the soil solution account for significant variation in plant health. It is not clear whether anyone has tried to quantify the standard deviation associated with variations in soil ecology or soil mineral proportion availability, but intuitively, it is easy to understand that such standard deviation would be extremely large. As explained earlier in this chapter, a healthy plant that has a high percent of sucrose in its sap is not vulnerable to insect damage. Thus, if the research study described above had been conducted in test plots of soil with diverse soil biology and proper mineral proportions in plant-available forms, the researchers might have measured a significant difference between the unsprayed control group and the experimental plants that were sprayed, with the control group having significantly better results. In this case, the insecticide might have damaged the healthy soil ecosystem and caused a deleterious result in the experimental group. The control group, though, would have produced healthier plants than had been seen before by virtue of growing in optimal growing conditions. But if the study were carried out on plants growing in lifeless soil deprived of nutrients, then plenty of pests would attack the plants in both the control and experimental groups. In that case the plants sprayed with insecticides might fare better than the unsprayed control plants.

Another example would be a side-by-side comparison between the performance of a genetically modified crop and an heirloom variety planted in a field that has little or no soil biology and is depleted of minerals in available forms. Researchers might see significantly different results from this study than from a comparison of the same two crops growing in a plot of biologically thriving soil with balanced mineral content.

Research reports seldom include any information about the health of the soil in which plants were grown. But an understanding of how plants grow in soils with optimal mineral content, a thriving ecology, and good energy flow needs to be established before the effects of fertilizers, pesticides, herbicides, or genetically modified crops can be realistically evaluated. The effect of these products on plants growing in unhealthy soils with unbalanced minerals should be cast in context and not assumed to reflect the results the products might create when plants are grown in healthy, minerally balanced soils.

Scientific literature that discusses the effects of fertilizers, pesticides, or herbicides or the performance of genetically modified crops seldom, if ever, takes this most important source of variation into consideration. It is unfortunate that most of the plant research data available is "petri dish science"; that is, not completely relevant to gaining insight into the workings of plants growing in biologically diverse, minerally balanced soil with proper energy flow.

Working with the New Garden Model

The new garden model leads to excellent results through its focus on using local materials and simple processes, on nurturing soil ecosystem diversity, on moving mineral proportions of the soil into balance, and on facilitating the flow of energy into soil. Knowing that healthy plants will naturally eliminate insect and pathogen pressures by virtue of the plants' own ability to thwart these challenges is empowering. The recipes in this book use local materials to produce broad-spectrum mineral amendments in forms plants can use directly. Because they are made from plants, these amendments by definition contain minerals in proportions that plants need and will respond to. The recipes for biological amendments use the full diversity of local biology to stimulate soils. In these circumstances, organic carbon becomes a limiting factor, and it is up to gardeners and farmers to supply enough organic carbon to feed the amazing diversity of biology in the soil. These activities produce other beneficial effects as well: carbon sequestration and a bountiful harvest of high-quality food.

Reaping the Harvest

The most pleasurable challenge for gardeners is keeping up with the harvest and bringing it to the table. Oh my! How many green beans can be eaten at one time! This a most wonderful problem to have, as long as there is enough time to do all the processing. Some crops, like beans, won't keep well in the garden, and there may be a lot of them that need picking. Using a dill pickle brine recipe, fill up ½-gallon jars with those extra beans and the brine for delicious eating all winter and even into the spring if the harvest is bountiful enough. Be sure to include garlic, a hot pepper perhaps, black peppercorns, a horseradish leaf or grape leaf to keep things crisp, and of course the dill that has self-sown throughout the garden. Try experimenting with other herbs, too. Pickling is a good storage method for any root crops. It is always a surprise to realize how good pickled vegetables taste during the winter months. And save that brine when the beans are gone. It makes a great addition to salad dressings, soup stock, potato salads, and more. I water the asparagus bed with the brine a couple of times a year. They like the salty, herb-flavored water as much as I do.

Dehydrating weeds and herbs all spring and summer long is an annual task much appreciated when the fresh supply of them is gone. Gallons of sealed glass jars are stashed away with stinging nettle, dandelion greens, tulsi, oregano, thyme, rose petals, bee balm, yarrow, elderberry flowers, and peppermint. When an abundant stand of mushrooms is found, try dehydrating the excess. When harvesting wild foods, always be sure that proper identification is confirmed by an expert source and always label the contents of those jars with name and date. Some mushrooms will dehydrate well if simply left in a paper bag in the refrigerator. Once the water is removed, the dehydrated material may be stored in a sealed jar for a year or more, a most efficient method of storage.

In the fall, after the plant energy has migrated to the roots for the winter, Joan and I put up root crops for the following year. We chop and dehydrate roots of dandelion, burdock, marshmallow, elecampane, astragalus, ashwaganda, and yarrow. Or we make tinctures using organic vodka.

Tinctures, tonics, elixirs, syrups, and shrubs are all storage tools that are fun to make, taste great, and will constantly remind us of the warmer days of summer during the cold months around the corner. It is tough to beat the flavor of homemade fire cider. That elderberry syrup will help prevent any cold or flu from entering a household.

Canning tomatoes should be a national pastime—perhaps it once was! Is there anything better than homemade tomato sauce in the dead of winter?

Berries are frozen on cookie sheets, then transferred to glass jars and put into the freezer. Blueberries, raspberries, peaches, and black currants are standard breakfast fare throughout the year. Eating high-quality berries—grown without the use of pesticides, herbicides, fungicides, or any other chemical additives—all year long brings a satisfaction that is difficult to explain. We dehydrate pears and apples because they have a lower water content. Jams and jellies require such a lot of sugar that we tend not to go that route, but they sure do taste good.

After the garlic is harvested, we cure it in the well-ventilated garden shed until September, and then clean and trim the bulbs, setting the largest aside for planting in October. The rest are put into canvas bags to be stored in the coldest closet in the house.

Potatoes may stay in the ground until the time is available to dig them up and store them. The longer they're left in the ground, however, the greater the threat of insect or vole pressure, so get to this chore when possible. Potatoes like it cold. A cold storage area is a good thing to have, just above freezing. Unlike a freezer, root cellars offer excellent storage at no additional cost than the initial construction materials and time.

And then there are carrots just left in the ground, mulched for the winter for us to steal when the ground allows, or savored in the spring after the ground thaws.

Whatever the garden provides, be sure to preserve the excess. Too much work has already been expended to let these opportunities go by. Try more than one storage strategy at the same time, to see which works better for a specific crop type. The rewards of eating clean food with little more effort than opening the freezer or opening a jar are well worth the while.

CHAPTER 2

Methods and Strategies

I t's time to take a deeper look at how gardeners can enhance the symbiotic relationship between soil and plants by applying home-made amendments made from local materials—the recipes in part 2 of this book. There are a few goals to pursue:

Building healthy soil. Soil is the result of eons of geological activity, the redistribution of minerals by plants as well as human factors, agricultural practices, air and water pollution, and more. As discussed in chapter 1, three key components of soil health are mineral proportions, biological diversity, and energy flow. Mineral proportions in forms plants can use define the elements needed to support photosynthesis, protein synthesis, and other biological functions. Biological diversity facilitates the digestion of the soil constituents into compounds the plant and soil ecosystem can utilize. Energy flow supports the level and distribution of nutrients throughout the eco-system. As with many living systems, simple improvements can lead to exponential improvements in the quality of the food you grow. It's a case of $1 + 1 = 10$.

Correcting deficiencies and excesses of mineral nutrients. Most soils are not optimized for growing food, although plants will grow just about anywhere. Consider that dandelion growing out of the crack in a sidewalk in the middle of a big city. Weeds grow in mineral-deficient soils, redistributing minerals from within to the surface. This is nature's mineralization mechanism. It is amazing to see the weeds in the garden change as the soil mineral proportions change. Balancing the mineral proportions in the soil provides the constituents needed for the optimal growth of cultivated plants. It will also dissuade the growth of weeds deemed undesirable.

Supplying the short-term needs of plants. It is empowering to have a shelf of homemade amendments that contain all the minerals that plants need for healthy growth. It also allows for precise response based on stage of growth. For example, a particular crop may need phosphorus during phases of its development. When a vinegar extraction of cow bones, with its large quantity of phosphorus and broad spectrum of other minerals all in plant-available forms, is on the shelf, it provides the opportunity for the gardener to play a vital part in the ecosystem, rather than simply being a bystander. You may find that changing from a more conventional approach that involves lots of purchased products to a sustainable, regenerative approach that relies on low-cost recipes made yourself using local materials takes more time and thought, but the benefits are many, and the time spent in the garden rather than driving to the store is far more thought provoking. Conventional gardening and farming practices can result in pollution, destruction of the soil ecosystem, and loss of soil carbon. It's inspiring to try new methods that do much to restore degraded soil, the local ecosystem, and the world at large.

Recording the steps taken in a garden notebook is extremely valuable in helping to recognize progress toward goals. As I get older, I can barely remember what I ate for dinner last night, never mind where I planted potatoes last year or how much rock dust of various types I spread in different parts of the garden or how many times and in what proportions. Being able to review what worked and what didn't, what was added or planted where and when, is helpful and fun. Documenting steps taken and results in a garden notebook for reflection in the following years will help improve future processes and methods. Every growing situation and set of available resources will be slightly different; good record keeping is the best way to capture these nuances.

Mineral Amendment Sources

Using packaged fertilizer products is a common way to supply deficient minerals, but these products often are manufactured in faraway locations, and their production includes hidden costs that are not reflected in the price tag. Some of the contents may be waste products from some manufacturing process, and they may include constituents not suitable for any garden. Mining of minerals has detrimental effects on the environment and the health of the people engaged in mining. There are packaging and transportation costs as well. It is well worth reading the labels of the mineral products sold for use in the garden. Do the listed ingredients add up to 100 percent? Are the contents agricultural-grade? All these factors motivated me to learn how to make my own mineral amendments, and to share what I have learned. It is wonderfully empowering to be able to make your own mineral amendments, especially from materials available in your own backyard or neighborhood at little or no cost! It also contributes to closing the waste gap, which is a quintessential sustainable, regenerative garden strategy.

Weeds and crop plants draw up minerals from beneath the soil surface, storing them in their leaves, stems, and seeds. Some weeds have taproots that can break through harder layers of dense soils, drawing

Stinging nettle is one of my favorite plants to include in amendments because of the broad spectrum of minerals it contains.

minerals from a depth of many feet; others have shallow roots that remain closer to the soil surface. Root depth has an effect on the resulting mineral uptake, as does the nature of the root exudates that plants release to stimulate specific types of soil biology. Each plant species has a different characteristic concentration and composition of minerals, which may not be an exact match with the mineral concentrations of the soil they are growing in. At the end of their annual life cycle, weeds die (annual weeds completely, perennial weeds only to the soil surface), and the minerals the weeds absorbed from within the soil are redistributed onto the soil surface. As a result, over a long span of time the topsoil's mineral composition changes, and the weeds that grow there will change also, moving the soil to nature's desired mineral composition. This is a constant evolutionary process that has produced rain forests, savannas, hardwood forests, and the other beautiful ecosystems on Earth. There's a lot we can learn by observing nature's model of soil mineral redistribution. As an example, specific types of weeds tend to grow in soils that have specific mineral deficiencies. By learning about the tendencies of such weeds, we can make educated guesses about which minerals are deficient in our soils. By choosing particular plants as the ingredients for the recipes in this book, it's possible to predict the specific concentrations of macro- and microminerals an amendment will contain, including trace minerals that are needed in amounts so small they are measured in parts per million (ppm). For more details about this, see "Weeds and Crop Plants" on page 113. Fermented plant juice is one example of a homemade mineral amendment. This process extracts the contents of a plant by fermenting the plant in organic sugar; the resulting extract contains other beneficial compounds besides minerals: enzymes and proteins, all in forms plants can use.

Plants send concentrations of essential nutrients to the fruit during reproductive growth. It's nature's method of ensuring that the fruit contains all that is needed to support the initial growth of next year's seedlings. Thus, end-of-season residual, damaged, or not-quite-ripe fruit may be the perfect ingredient to produce an amendment to feed that same crop next year. Ferment those damaged, insect-eaten, or not-quite-ripe peaches this season to feed next year's peach trees. Ferment those end-of-the-year tomatoes to make a shelf-stable amendment to feed next year's tomatoes, and so on. Combining leaf mold, rainwater,

Table 2.1. Mineral and Biological Amendments at a Glance

Recipe Name	Type of Amendment	Comment	Page Number
Water Extractions	Biological, then mineral	Easy to make, but quickly starts to smell very bad!	122
Apple Cider Vinegar	Weak acid for vinegar extractions	Easy to make, so why buy vinegar?	125
Vinegar Extractions	Mineral	Broad-spectrum mineral content. Close waste gaps by making this recipe.	129
Fermented Plant Juice	Mineral	Broad-spectrum mineral content in plant-specific proportions.	134
Fermented Fish	Mineral	Broad-spectrum mineral content. Close a waste gap by making this recipe.	140
Leaf Mold Fermentation	Biological and mineral	This recipe defines sustainability: rainwater, weeds, and leaf mold!	144
Leaf Mold Biology	Biological, then mineral	The quintessential local biology source.	149
Lactic Acid Bacteria	Biological	Refrigerator-stable biology source.	154
IMO #1	Biological	Mainly used for making IMO #2.	158
IMO #2	Biological	Refrigerator-stable local biology source.	164
IMO #3	Biological and mineral	Used to make IMO #4. Animals like to eat this as it ferments.	167
IMO #4	Biological and mineral	Local biology digests soil and rock dusts to create a customized, living mineral amendment.	173

and damaged fruits may be the most sustainable, low-cost amendment process in the world!

Another amendment process is a vinegar extraction. Organic apple cider vinegar may be used to extract minerals from shells or bones. The weak acid in vinegar breaks down minerals into water-soluble forms. Vinegar extractions of eggshells will produce a concentration of calcium; bone extractions will produce concentrations of calcium and phosphorus. Vinegar extractions are shelf-stable, water-soluble, broad-spectrum mineral amendments that can be absorbed directly by plants as foliar sprays or the soil/root ecosystem as a drench. It's even more satisfying and sustainable to make apple cider vinegar using collected rainwater and apples that dropped from a local tree. (See the "Apple Cider Vinegar" and "Vinegar Extractions" recipes in chapter 6.) Keep in mind that these recipes are forgiving and the proportions need not be followed exactly. Indeed, I encourage experimentation. For instance, try repeating the vinegar extraction process several times using the same batch of shells or bones to get the most out of them. Performing repeated vinegar extractions of the residues left over after making fermented plant juice can result in different concentrations of minerals in each subsequent extraction. Most of the time the concentrations of minerals will decrease, but in some cases the concentrations increase in the later extraction.

Other local sources of macrominerals are quarries that mine rock deposits or silt or clay along riverbanks. Another source of minerals is the waste from a local restaurant. The shells from oysters, clams, mussels, and other shellfish are usually just thrown into the garbage. Ask if they could be tossed into a bucket to be picked up when convenient. This is an excellent way to close a waste gap. For those who live close to the ocean or sea, mineral-rich materials that wash up onto the shore, such as seaweeds and marine shells, are excellent sources of minerals. Seawater is the quintessential mineral source; it contains all the macro- and trace minerals needed in the garden. The key is to recognize a new mineral amendment resource when you come across it, whether it is a pile of silt, oyster shells, or a patch of nettle, and to utilize these resources as availability and time permit. Closing the waste gap is sustainable and regenerative. For details about using these materials as amendments, see chapter 5.

Biological Amendment Sources

The idea of "fertilizing" a garden with biological amendments may seem strange, but introducing vast quantities of local biology into a soil increases digestion of the soil constituents and provides plants with increased access to water and nutrition. Introducing this biological diversity is a most important aspect of improving the quality and health of the soil ecosystem. The redundancy of organisms to complete tasks within the soil establishes resiliency and stability, and also keeps pathogens in check. Plant roots then come in contact with a more diverse selection of bacteria, fungi, and archaea, selecting and feeding those that will supply nutrients improving the plants' ability to support their individual functional needs. Any discussion about "good" or "bad" biology is fraught with unknowns.

Local Biology

The biology needed to make local, sustainable amendments is all around. The floor of the woods contains billions of microbes. The leaves, stems, and roots of all plants are coated with biology. Harnessing these local biology sources to make agriculture amendments is free to all. And here is a paradigm shift: A large biological diversity competing for available resources keeps in check pathogenic biology that competes for those same resources. (This phenomenon holds true in human society, too: Increased diversity keeps bad influences at bay.) A healthy soil ecology with balanced mineral content is able to provide plants with the compounds needed to build defenses and increase resistance to pests, pathogens, drought, and other stresses. It is only when some part of this ecology becomes unbalanced that soilborne pathogens thrive, and this is as nature intends. Unwanted pathogens do not suddenly arrive in a garden bed to destroy a crop—they are always present there. However, the rich diversity of a healthy soil ecology usually keeps them under control. Rather than add a fungicide to kill a target pathogen, which may deleteriously affect other aspects of the soil ecosystem in unknown ways, it is more effective to add vast numbers and varieties of local biology to the ecosystem. Doing so will require the pathogens to compete for available resources to stay alive, but it will not eliminate them altogether. As all these aspects of a soil ecology come together, the increase in plant health, plant resistance to pathogens and insect pressures, number and diversity of pollinators, and the diversity of the entire aboveground ecosystem become exponential.

One of the best ingredients to use in biological amendments is leaf mold, which is found on the floor of any wooded area and defines the local diversity of biology in the soil for that time of year, at that elevation, and at that temperature. Many thousands of varieties of bacteria, fungi, and archaea are represented in leaf mold—some aerobic, some anaerobic, some facultative.

It's common practice these days to buy biological inoculants that come in packages and are advertised to contain several kinds of bacteria and several kinds of fungi in dormant forms. But is there any assurance that these biological inoculants will come alive and thrive in their new soil home? How could one expect a handful of species to be the best choice for gardens situated in different regions and environmental conditions? Not only is the local soil biology of every garden unique, but the species composition changes as the seasons change and soil temperatures change. The biology propagated in the spring will be different from that propagated in the fall. The biological amendment recipes in this book harness this local diversity, and I believe they will provide significantly different results than purchased inoculants can. It seems foolish to suggest that a handful of microbial strains in a commercial inoculant could be as effective as the diversity and abundance of local varieties. And, of course, those local varieties are available at low or no cost.

The use of these biological amendments creates a tremendous need for organic carbon, because carbon is a basic building block of cells and carbon compounds are a critical source of energy for all living things. The soil biology digests available organic carbon in order to fuel the growth of its own populations. Carbon-eating microbes are then eaten by larger predators, and the waste products and decaying bodies become nutrients within the soil solution. Mulches, compost, and cover crop residue are quickly decomposed, and the organic carbon in them is sequestered into the soil, increasing stable forms of humus and the soil's exchange capacity. Balance is attained by the interaction of these biologies and their interaction with the plants whose root exudates selectively feed the soil biology.

It is amazing to realize that the soil needs as much combined carbon as possible to feed a thriving soil biology rather than the expensive nitrogen fertilizers that gardeners and farmers are led to believe are essential for success, but that often end up leaching into nearby watersheds. With soil mineral proportions in the right range, air freely flows through the soil. With soil biology in full gear, nitrogen-fixing biology will provide the plants all the nitrogen they need. These concepts also explain why the loss of topsoil in the Midwest and around the world has contributed so much to global warming. The loss of topsoil and subsequent release of combined carbon into the atmosphere attributed to farm plowing and tilling practices are a major contributor to worldwide carbon emissions. Feeding the biology in the soil also provides the large-scale environmental benefit of sequestering carbon. Yes, this all feels right!

When a crust forms on top of a thick layer of mulch such that water runs off rather than being absorbed, adding these biological amendments can solve the problem. The biology digests the organic carbon in the top crust and then throughout the mulch, quickly changing the compacted mulch into airy soil that will store water and allow air circulation, which in turn will support the soil ecosystem. Add biological amendments to your compost pile to facilitate decomposition. Fungi, bacteria, and archaea each digest different aspects of the compost pile. Fungi decompose lignins, for instance, while bacteria decompose cellulose and chitin.

Shelf-stable biological amendments are valuable because they are ready for use whenever you need them. Some recipes can be made only when environmental conditions are right. In New England, for example, leaf mold biology probably won't be a practical choice to apply in January, because local sources of leaf mold will be frozen or buried under snow. Lactic acid bacteria, however, can be made any time of year and may be stored in the refrigerator for a long time. Indigenous microorganism number 4 (IMO #4), which I describe below, is another example of a stable biological amendment. It is also a mineral amendment. Making a pile of IMO #4 in the fall provides a good supply of active biology in early spring when leaf mold biology would be difficult to make. Non-shelf-stable biological amendments include leaf mold biology, water extractions, and raw milk.

There is a relationship between a thriving soil biology, which provides complex compounds to the plant root system, enabling the manufacture of secondary metabolites, the subsequent resistance to insects and pathogens, and the ability to measure this with a refractometer. Plants grown using the new garden model also produce high-order compounds needed for human health. They store longer and taste significantly different from plants grown using a simple plant model.

Indigenous Microorganisms (IMO)

I first encountered descriptions of the making of IMO in the book *Natural Farming Agriculture Materials* by Cho Ju-Young. This is a four-step process, and the four steps are named IMO #1, IMO #2, IMO #3, and IMO #4. The process first captures local biology from the woods (IMO #1) and ferments the biology into a refrigerator-stable form (IMO #2) that can then be cultured to make IMO #3 on an as-needed basis. Once IMO #3 is alive and thriving, additional materials may be added and digested to produce IMO #4. The entire process takes about a month to complete, with each step needing about a week to "ripen."

IMO #2 is shelf-stable when stored in a refrigerator and may be used as a biological amendment added to the soil or compost pile. IMO #3 and IMO #4 may be understood as two different stages of digestion. IMO #3 is the inoculation phase that initiates biological activity within a substrate (organic wheat bran), and IMO #4 is a fully inoculated material capable of digesting minerals like rock dusts, shells, and soil.

The mineral composition of the final product (IMO #4) can be customized in accordance with a soil's needs (as interpreted from soil test results) by adding liquid mineral-rich amendments during the IMO #3 and IMO #4 phases and rock dusts or other "hard" mineral sources during the IMO #4 phase. In addition to being an extremely effective mineral and biological amendment, IMO #4 has many further uses on farms. It will digest the anaerobic waste products of chickens or pigs to maintain a clean, odor-free environment. IMO #4 is a starting point for experimentation. Because it is a living, mineral-rich product, it may be used to digest other materials found locally, and to augment the mineral contents and biological activities of the compost pile. When IMO #4 is left sitting on the ground in a shaded spot, the biology in it will go dormant, ready to come alive when applied to the soil, mulch, compost, or other organic substrate. Thus, it is a versatile, stable amendment that can be manufactured and used throughout the growing season. Note that the biology used to make IMO #1 will be peculiar to the location and time of year of its capture, so the IMO #2, #3, and #4 will also have these characteristics.

Amendment Strategies

It takes a year for the Earth to travel one full revolution around the sun. During that time, the soil undergoes cycles of change. The intensity of soil cycles at any given location is related to its latitude—its distance from the equator. Freeze-thaw cycles and rainy season–dry season cycles are examples. These annual cycles present timely opportunities to improve a soil's balance of minerals, biological diversity, tilth, humus, electrical conductivity, and other properties of interest.

Developing a soil amendment strategy builds from the new understanding of the importance of diverse soil biology, balanced broad-spectrum mineralization, improved energy flow, and increasing the soil exchange capacity as aspects of improving soil ecosystem health. Improving the nutrition of the crops grown on a specific piece of land takes time. After the first season of working to improve very poor soil, better crop performance may be readily noticeable. But as the soil continues to improve, it becomes harder to achieve better results. It is always the weakest link in the system that defines plant health, and it becomes harder to identify and act on that weakest link. Some amendment strategies will include activities that produce an effect quickly, but other activities will take years to produce results. Feeding plants using foliar sprays on a regular basis throughout the season is an example of a short-term annual activity. Mineral proportions, humus content, and exchange capacity are examples of soil characteristics that may take years to optimize. It may take many small applications of a specific amendment over the course of several years in order to fix macromineral deficiencies or excesses.

All soil improvement activities are deliberate, important, and part of the annual cycle of garden activities. These are endlessly satisfying activities, being a part of the ecosystem, stimulating the vitality of the

Soil Improvement Activities

Here are some activities that can be conducted throughout the growing season. These tasks will quickly become a way of life. Some may be done only once every few years, like a soil test. Others are seasonal, like starting seeds, still others can be added to the list of what to do today in the garden as the opportunity arises, like harvesting nettles to make an amendment. I offer these lists as a starting point. Customize them to match your garden's unique conditions and the amount of your time available for gardening.

ACTIVITIES FOR LONG-TERM IMPROVEMENT

- Conduct a soil test in the fall.
- Consult the soil test results to establish an amendment mixture that will address macromineral deficiencies and soil energy needs.
- Amend the soil with sources of macrominerals (rock dusts, crushed shells, and compost) and organic carbon (mulch or cover crops) in the fall, winter, and early spring.
- Apply biological amendments, compost, fermented plant juices, and mineral extractions to the base of perennials in the fall and spring.
- Foliar-spray fruit trees with broad-spectrum mineral amendments in the fall before the leaves drop, to the bark and new leaves in early spring.
- Replant a second round of a crop, sow a cover crop, or mulch garden spaces after harvest rather than leaving soil bare.
- Make new compost piles annually for use in following years.
- Make IMO #4 in early spring and early fall. (It's good to have on hand throughout the year.)
- Gather leaves, use a lawn mower or shredder to shred them, and distribute on the garden.
- Start a new garden (at any time of year).

ACTIVITIES FOR SHORT-TERM IMPROVEMENT

- Monitor soil moisture and plant appearance and water plants as needed.
- Apply foliar sprays on a regular schedule (such as every 7 to 10 days) throughout the growing season.

▸ Apply biological and mineral amendments at any time of year directly to the soil as a drench or in conjunction with mulching.

▸ Apply a mixture of biological and mineral amendments, compost, and small amounts of a rock dust mixture at or before planting or transplanting.

▸ Cover crop or mulch garden spaces after harvest rather than leaving soil bare.

▸ Make shelf-stable amendments anytime resources become available.

▸ Make non-shelf-stable amendments as available or needed.

▸ Keep an eye out for potential sources of minerals, carbon, or biology for making amendments (but not while driving, unless you're in the passenger seat).

▸ Use a refractometer to assess whether plants reacted positively or negatively to foliar spray applications.

▸ Use a refractometer to assess whether boron is needed.

▸ Harvest, eat, or store weeds and other wild plants as they come into season, such as nettles, purslane, dandelion leaves, chickweed, autumn olive berries, barberry roots, dandelion roots, and acorns.

▸ Gather and store food for the year to come as it becomes available (root cellar, ferment, dehydrate, freeze, can).

▸ Measure soil temperatures to gage when to start seeds or transplant.

landscape, and producing healthy crops. Improving the health of the environment and ourselves every year is a wonderful way of life.

Start with a Soil Test

A soil test is a good place to begin developing an amendment strategy. A soil test will provide an indication of the mineral proportions within the top 6 inches (15 cm) of the soil, the amount of organic matter in the soil, the soil exchange capacity, and the soil's overall mineral needs and excesses. These results will be the starting point of calculations to determine what macrominerals are needed to be acquired and how to create a custom macromineral amendment recipe. This custom recipe may then be added to the entire landscape as part of the long- and short-term mineralization strategy.

Calcium, sulfur, boron, silicon, manganese, basalt, and granite rock dusts, along with crushed shells or bones, are all candidates to consider for the blend. The rock dust from local lime quarries contains calcium; the rock dust from basalt quarries contains silicon and perhaps manganese, and may be paramagnetic; shells and bones will contain calcium, phosphorus, and trace minerals. Borax is an easy source of boron (it contains roughly 10 percent boron) and is a rock dust mined in the western United States. Gypsum is a source of calcium and sulfur ($CaSO_4$-$2H_2O$); diatomaceous earth is a source of silicon. What are the resources in your backyard that you may not have thought of in the past? It may take several years to bring the mineral proportions of your soil to the right levels, depending on the initial degree of deficiency or excess. Chapter 4 goes into detail about how to interpret soil test results and use them to calculate specific amounts of nutrients to apply. The "Local Rocks and Soils" section on page 116 provides more detail on sourcing local rock dusts.

Soil test information can form the basis for both short- and long-term amendment program strategies. Use soil test results as target goals, guides to improve the soil ecosystem year after year. Keep in mind that if you're working with soil where conventional fertilizers have been applied in the past, your soil may actually have developed some mineral excesses. So a logical, and moneysaving, part of your strategy will be to ensure that you are not adding more of those minerals, allowing the soil time to come back into balance. After a few years of implementing your long- and short-term amendment strategies, conduct another soil test to see what changes have occurred.

Apply Foliar Sprays and Drenches

Another good activity to establish right away is a regular plan of foliar spraying and drenching; this is an integral part of a short-term amendment program. The recipes in this book provide a broad spectrum of minerals in forms that plants can use, and applying them to plants as sprays or drenches is a quick way to respond to known soil mineral deficiencies and signs of plant stress. When potato beetles appear on the potato plants, a broad-spectrum amendment foliar spray will provide the plant with the nutrition it needs to increase the sugar content of its sap and repel those beetles. (Check water needs as well.) When no signs of plant stress are apparent, be proactive by applying broad-spectrum foliar sprays to feed the plant. This will eliminate the risk of deficiencies throughout the entire growing season.

Apply Dry Minerals

Late fall, early winter, and late winter are the best times to apply dry minerals that require digestion by the soil ecosystem. These sources of macrominerals include rock dusts, crushed shells, silts, and clays. (For more details on these materials, see "Local Rocks and Soils" on page 116.) The goal is to increase dispersion and facilitate the decomposition of these hard materials into the soil solution. Applying small amounts of macrominerals two or three times per season is more desirable than applying all at one time. It's also a good practice to apply them in conjunction with a mulch, cover crop, and biology and to time applications when rain or snow is anticipated to soak into the ground. Subsequent soil test results will help in evaluating the results of these efforts.

Keep the Soil Covered

A critical aspect of both short-term and long-term strategies are to *never* leave the soil bare. Remember that healthy plants send about 25 percent of the sugars they make during photosynthesis downward through the phloem into the soil to feed the soil biology. Thus maximizing the amount of time that growing plants are present in a garden or field helps keep soil biology thriving. There are many ways to achieve this goal:

- As soon as one crop is harvested, plant another food crop, or a cover crop.
- Learn to see weeds as a most important garden resource. You can influence what kind of "weed" cover will fill your garden beds in the future by strategically allowing certain types of plants to go to seed. A variety of food crops are great candidates for this approach—see "Weeds" on page 91 for details on managing food crops in this manner.

- Let the local weeds grow as a cover crop, but don't allow them to go to seed unless you want them back. Once they begin to flower, cut them down at the soil surface, leaving the roots in the ground. Use the cut plant tops to cover the soil surface as a mulch.
- Sow cover crop seed in early fall or even later as your growing season permits. A cover crop that dies back during the winter becomes a mulch; the roots nourish the soil.
- Mulch unplanted areas with crushed leaves, hay, or straw. The organic carbon in the mulch feeds the soil biology and will also increase the soil's exchange capacity.

Apply Biology

Applying biological amendments facilitates digestion of leaves, straw, hay, bark mulch, wood chips, or any form of carbon and stimulates the soil ecosystem. IMO #2, IMO #3, and IMO #4, as well as lactic acid bacteria, raw milk, or leaf mold biology, may be added to stimulate the soil biology in early autumn and early spring. Remember that the locally grown biology is the biology that wants to thrive in your garden!

A Quick-Start Guide

Many gardeners I talk to are somewhat skeptical about the idea of committing to a full-fledged amendment strategy. They feel as though their gardens are doing pretty well already. But what many gardeners don't realize is how much *better* food could taste when crop plants experience optimal nutrition. Later in the book, I explain the Brix scale developed by Dr. Carey Reams for evaluating the quality of fruits and vegetables. Dr. Reams's work shows a Brix value range for tomatoes from 4 to 12, but most people have never tasted a tomato with a quality level that exceeds 5. We have no concept of what that Brix value 12 tomato would taste like, or what the health benefits of eating such nutrient-dense tomatoes might be! Flavor and nutrient density are what motivate me to follow my amendment strategy, and the results are well worth the time I invest.

If the idea of developing your own amendment strategy seems overwhelming, try following this quick-start guide to using amendments. Begin with step one, and build from there as time allows and your confidence grows:

1. **Apply leaf mold biology.** Applying leaf mold to the garden in the spring is the easiest, cheapest, best way to start improving your garden results. Apply it again a second time later in the season, too.

2. **Make one or two shelf-stable mineral amendments.** Once you have amendments in jars waiting on a shelf in your basement or pantry, you're much more likely to actually mix and apply a foliar spray or soil drench. Try the recipe for Vinegar Extractions or Fermented Plant Juice—both are easy and versatile.

3. **Add extra minerals and biology to the compost pile.** Use the digestive capabilities of a compost pile to break down macro-minerals into forms the garden can use. Those kitchen scraps have to go somewhere! Add other ingredients such as rock dusts and biological amendments throughout the year as they become available. Apply finished compost to everything you plant to provide minerals, biology, and humus.

4. **Apply mulch and grow cover crops.** This can be as simple as allowing the weeds to grow in a bed and then cutting them down for mulch. Or you can selectively allow some of those weeds to go to seed, and then collect your own "cover crop" seed for the next season.

5. **Develop a foliar spraying routine.** Spraying seems like no fun because of its association with pesticides and other harmful products. But when you use foliar sprays to apply homemade amendments, the results will change the way you think. Foliar spraying is an amazingly effective way to provide plants with the nutrients needed throughout the entire growing season. A regular schedule of foliar spraying (once every 7 to 10 days) will significantly increase the quality of the plants grown in any soil.

Starting a New Garden Plot

The new garden model recognizes that the existing soil ecosystem growing in a future garden site is valuable. Perhaps this ecosystem may be harnessed and a gentle transition to a revised ecosystem can occur. Suppose the local biology is allowed to transform the grass into seed-ready soil over a period of time. Minerals may be added as time permits, biology added to facilitate the transition when needed. All of this will take months, but the work will be accomplished by nature, requiring only brief moments of effort by the garden maker to encourage the process. This is a great introduction to what I call garden time and the recognition that there is a constant flow of change, some processes take time to unfold, and we are all a part of the flow of the universe.

There are variations on this approach to starting a new garden. The simplest version is to mulch thickly and wait patiently while nature prepares the area for planting. Cover the future garden area with a layer of crushed leaves, straw, or hay 9 to 12 inches (23–30 cm) deep. You can use a combination of these materials, too, or whatever similar materials you have available. Let the area sit for 6 months or even longer.

We love our perennial garden for its beauty and color. It shelters and feeds beneficial insects, and some of the plants, such as these coneflowers, provide us with good medicine, too.

During that time, the biology under the mulch will have degraded the mat of grass and thick grass roots enough to be ready for planting. All you need to do is pull back the mulch, loosen a small area of soil underneath, and plant something. Keep the mulch intact to continue the garden-making process and suppress weeds from taking root.

Another version is to use the occasion of establishing a new garden area as an opportunity to immediately put an amendment plan into practice. This strategy can be very effective. Establish a new garden area to be amended. Apply rock dusts, silts, and/or crushed shells as needed and cover them with 9 to 12 inches (23–30 cm) of mulch. Then thoroughly water the surface of the mulch and apply a biological amendment. If possible, time this activity for just before a rainy day, which then complements the watering activity. The biology will digest the organic carbon in the mulch quickly, breaking it down into humus, which will increase the exchange capacity for next year's crop. You can also apply a foliar spray to the surface of the mulch to increase the trace minerals that invariably will also be needed. Consult soil test results if available to guide your choice of amendments. If these are not available, use a broad-spectrum amendment such as a nettle or dandelion extraction. This is a great way to start a new garden in the fall directly on areas of lawn or other grasses. And it is so much less daunting than the work required to start a new garden in the conventional model. That model requires the soil to be rototilled or plowed to eliminate plant growth and clear the way for sowing seeds. Next, that plant material is allowed to dry out so it can be raked out of the garden area and discarded. It's then easy to plant seeds in the dry, loose, weed-free soil, but the soil ecosystem has been significantly damaged. The delicate strands of fungal mycelia have all been destroyed, and the dried-out soil no longer holds the water that would support the soil biology needed to expand the reach for water and nutrients of the plant root system. In short, significant time and energy have been spent to prepare a less-than-optimal soil.

A more nuanced approach to starting a new garden bed includes additional steps that will further improve the result. In this case, before spreading the layer of thick mulch, conduct a soil test in order to gain information about existing mineral proportions. With this data in hand, it's possible to devise a more specific strategy to address mineral deficiencies and excesses. For example, if there are gross deficiencies of macrominerals, spread a specific mixture of rock dusts, silts, and/or crushed shells over the plot before applying the thick mulch and then again over the surface of the mulch several times during the months when the biology is decomposing the sod. To supply microminerals that are identified as deficient, apply a drench that includes a specific fermented plant juice or vinegar extraction to the surface of the mulch layer and again at intervals for the next several months.

Add biology initially to facilitate decomposition of the organic carbon mulch into humus and to increase the soil's ability to hold on to mineral ions (the soil exchange capacity). Leaf mold biology is a cheap, easy-to-make product for this purpose, but difficult to make in the winter here in New England. A water extraction of freshly picked weeds is a cheap and easy biological solution to apply from time to time. Refrigerator-stable lactic acid bacteria and raw milk are other options. IMO #4 provides both biology and minerals and is a highly effective product to spread on top of the mulch of a new garden bed. All these biological amendments will facilitate the decomposition of the mulch and the underlying sod root structure.

Additional steps may be incorporated as time and resources permit, such as topping the surface of the thick mulch with a thin layer of soil or compost or adding extra biology and mineral applications based on soil test results. The new garden will be ready for planting next season, and the time from now until then may be used to nurture that space. The machinery required is the soil biology and the time required is the months for the biology to do the work. I have used this practice repeatedly and planted potatoes, garlic, and other crops on what was thick grass after 6 months of digestion.

Amending at Planting Time

Another very effective strategy is to view every crop planted as an opportunity to improve the health of the soil for the next season. Add compost, small amounts of a rock dust blend, mineral amendments, and local biology to stimulate the soil ecosystem, and remove rocks whenever planting anything. This is a regenerative philosophy of gardening. A regenerative methodology may require a greater investment of time when planting a crop, but the soil improvement, the health of the plants, and the long-term results are well worth it for the people who will eat the crop and for the environment. This approach to planting will also result in greater ease when planting crops in the future. A regenerative approach is especially important in the early years of garden development or when an old farm or garden space is being rejuvenated.

Regenerative Potato Growing

Growing potatoes can be an excellent example of a regenerative growing process. Of course, there are many ways to grow potatoes. You can simply place seed potatoes on the soil and cover them up with a thick mulch of straw or hay, and potato plants will grow and form a crop of tubers. However, this way of growing potatoes will not significantly improve the soil structure, mineral content, biological activity, or depth of tilth.

Consider instead a potato planting system that improves the soil. The first step is to prepare the bed for planting by scraping the soil surface with a broad hoe to remove any existing plant tops. Work gently so as not to disturb the soil structure below the surface. The goal is to protect the soil ecosystem as much as possible. Put the green plant residues into a compost pile or spread them as a mulch somewhere in the garden.

Next, gather ingredients for planting: a wheelbarrow of sifted, aged compost; a rock dust mixture that represents your garden's mineral needs; and some IMO #4. (See "IMO #4" on page 173.)

Prepare a solution to soak the seed potatoes in prior to planting. Use rainwater and add a broad-spectrum mineral amendment like fermented plant juice of dandelion or stinging nettle or leaf-mold-fermented carrot tops. A calcium source like vinegar-extracted eggshells or oyster shells may be included; or use extracted cow bones if phosphorus is also wanted. Also add a biological amendment like leaf mold biology, lactic acid bacteria, or spoiled raw milk. Mix all of this well, using the technique described in "Good Water" on page 74. This kind of solution provides mineral nutrition and biology that the new potato sprouts can utilize right away. If the specific ingredients suggested here are not available to you, substitute what you have available.

And even if you don't have mineral or biological amendments ready, it's still helpful to soak the seed potatoes in plain water. I use whole seed potatoes, and I do not cut them into pieces. I soak 7 to 10 potatoes at a time. As I prepare each planting hole, in goes a potato until all are in the ground. Then I start the process over again.

For each seed potato, dig a hole about 12 inches (30 cm) in diameter and about 12 inches deep. Put any rocks encountered into a bucket, crumble the soil by rubbing handfuls together, and place the loose soil next to the hole. I used to toss the removed rocks into the woods, but quickly learned that this material would make a great stone wall. Depending on your soil's composition, you may find that you can collect enough rocks for a wall rather quickly. Next, put a couple of handfuls of aged compost into the hole, and add a light dusting of rock dust mixture, a very light dusting of IMO #4, and enough of the loose soil to fill the hole about half full. Mix all the ingredients completely by rubbing handfuls together until the soil in the hole is uniform and crumbly.

Place the soaked potato an inch or two (2.5–5 cm) from the bottom of the hole and cover with the rest of the amended soil in the hole. This will fill a little more than half the hole. Water the potato with good water mixed with a balanced mineral amendment like nettle-fermented plant juice or leaf-mold-fermented carrot tops. Use the soil that remains next to each hole to fill that hole as the potato plant grows. This soil may be removed from the potato plot and further improved by mixing with compost, and the entire area can be mulched with straw, hay, or chopped-up leaves to retain water. This amended soil will then be used to fill the remaining space in the hole as the potato grows out of it. Eventually the hole will be filled with a growing potato plant and a thriving soil. Mulch any remaining space around each potato. Pay attention to the need to water if the soil surface is left exposed to the air. Once the holes have been filled, be sure to mulch the bed if not done already. It may take 2 to 3 weeks for the potato sprouts to emerge out of their new home.

Space planting holes about 16 inches (40 cm) apart in a row, with rows about 16 inches apart. Every third row, leave a little more space between rows as a walkway. This spacing allows the soil structure to be maintained to some degree.

Be sure to keep the area hydrated by watering as needed during any period of time when the soil is exposed to the sun and dry air. How much and how often will depend on local environmental conditions. Regular light watering in the morning will keep soil moisture levels consistent; it also enables assessment of whether any further watering is needed. There is no substitute for regular observation. Once the area is mulched, the need to water will return to normal. After leaves have appeared apply a broad-spectrum mineral amendment foliar spray once every 7 to 10 days.

At harvesttime a similar soil enrichment process is invoked. Gather materials: a wheelbarrow of sifted, aged compost, a rock dust mixture that represents the garden's mineral needs, a wheelbarrow of a mineral-rich plant like stinging nettle, and some IMO #4. I have a stand of stinging nettles reserved for this purpose that I use each year. I chop it up with clippers, filling several wheelbarrows for the purpose of feeding the soil. Be cautious about handling this plant. It just may sting.

Dig up each potato plant in turn, trying to find every tuber large and small. Deposit the remains of the potato plant, a couple of handfuls of compost, a couple of handfuls of stinging nettles, a light dusting of rock dust mixture, a very light dusting of IMO #4, and some of the removed soil into the hole. Mix all the components within the hole, striving for even distribution of the contents. Cover with the remaining removed soil and move on to the next hole. When this task is completed, cover the entire plot with a rock dust mixture, mulch it, sow a cover crop, or all of the above. Sprinkle IMO #4 on the surface of the mulch, and water with a biologically rich solution of leaf mold biology, lactic acid bacteria, raw milk, or whatever is available. If none of these biological sources are around, make a quick water extraction using any plants available to supply some biology to the surface. (See "Water Extractions" on page 122.)

By the following growing season, the soil's tilth, depth of tilth, mineral proportions, exchange capacity, biological diversity, and structure will all be significantly improved. By employing this technique, the planting of potatoes becomes a fundamental garden improvement program as you systematically change the location of the potato plot in your garden from year to year. Eventually, your entire garden will have undergone this radical transformation. The first time you use this method to prepare a plot may be quite time consuming, but subsequent rounds will proceed more quickly and easily as the soil structure improves. I have harvested 17 tubers per plant using this technique as well as significantly improving my soil structure.

I employ these planting techniques with everything I plant. Before planting, I always mix two-year-old compost, my standard rock dust mineral mixture, and IMO #4 into the existing soil. I apply mineral amendments as foliar sprays throughout the growing season as well. Harvest is another opportunity to further amend the soil. The principles remain the same, and the results are very satisfying.

Amending When Transplanting

Planting transplants is a convenient way to establish a crop quickly, prevent insect damage to those tasty, succulent new shoots, and get a jump on the growing season. I've learned that transplanting may be done at any time of day, even when conditions are hot and sunny. The key is proper root-ball hydration to ensure a continuous flow of water through the xylem pathway so that the plant does not dry out and become stressed. This is accomplished by soaking the soil *and* the roots prior to putting the plant into the ground.

The first step is to prepare the soil. Dig a hole roughly twice the width and a little deeper than the pot the transplant is in. Pick out the rocks from the removed soil and crumble the soil by rubbing handfuls together. Next, toss a handful or two of well-aged, sifted compost into the hole. (This completely composted material contains humus, which will improve the exchange capacity of the soil.) A dusting of a custom rock dust recipe may also be added, along with a very small sprinkle of IMO #4. (Go easy on the IMO #4, as this material is very biologically active.) Mix these materials into the soil by again rubbing handfuls together. The goal is to create a crumbly soil texture for the plant roots to burst into. Also prepare a bucket of water that includes mineral amendments and biology, as described above in "Regenerative Potato Planting."

Once the hole and the water mixture are ready, lower the transplant, pot and all, into the water. Do not let the water spill over the top edge of the pot. Hold the pot so that its top edge is just above the waterline and allow water to infiltrate the soil through the drainage holes at the bottom. Wait until water appears on the surface of the soil

Transplants, Temperature, and Trends

Monitoring soil temperature at transplant time is important for optimal plant growth. In the Northeast, planting tomatoes in early May when the soil temperatures are still in the range of 50°F (10°C) can be disappointing. Plants just sit there, making no new growth until the soil warms up. I watch for the appearance of volunteer tomato seedlings as a cue for planting. When those self-seeding tomatoes come up, I know that the soil is warm enough to transplant. By measuring the soil temperature around those volunteer tomato seedlings, I gain an understanding of when to plant in subsequent years. Many seed catalogs provide information on optimal temperature ranges for germination as well. I keep track of lots of information like this in my garden notebook for future reference.

I have been measuring the soil temperature in my gardens (situated at about 42 degrees latitude) every day for many years now and have noted disturbing trends. I like to record the lowest soil temperature of the day in the early morning before the first rays of the sun begin shining directly on the ground. The occurrence of low soil temperatures that are greater than 70°F (21°C) has increased considerably in the past seven years in my gardens. The average number of days low soil temperatures exceeded 70°F for the years 2013 through 2017 was 9 days. But in 2018 I recorded over 30 with the highest daily low temperature at 74°F (23°C). During the day the soil temperature may rise 10 to 15°F (6–8°C) above the morning's low temperature. With our fundamental lack of understanding of the soil biological ecosystem, it is difficult for me to imagine how these high soil temperatures might affect the soil biology and the plants that grow in them. Some plant physiological functions cease to occur at increased temperatures. Higher soil temperatures will affect evaporation rates. And when soil temperatures are continually so warm, it is likely that gardens where the soil surface is left bare will not stand a chance of supporting either soil biology or crops. Mulch or cover crops will help hold water in the soil and reduce soil temperatures, enabling some respite from these challenges.

in the pot. This confirms that the entire root ball is saturated and ready for its new home. Next, cup one hand around the transplant stem and tip the pot upside down. With your other hand, tap the bottom of the pot to release the root ball from the pot. Set the pot aside and turn the transplant right-side up again. Rest the root ball in the planting hole and gather the soil around it. Tamp the soil around the plant lightly to verify that the root ball is secure. The level of the soil surface should be the same as the surface of the root ball when it was in the pot for most plants, but not tomatoes, for instance. Some gardeners cover much of the tomato stem with soil when planting, as well as the root ball.

Repeat this process for each transplant in turn. When you have set all the plants in place, use the remaining amended water to soak the soil around the plants completely. Mulch the surface with about an inch (2.5 cm) of straw, hay, crushed leaves, or cut grass depending on the strategy for this area during the remainder of the growing season. A thin grass mulch will not last as long as a thick leaf mulch, for instance. Be sure to cover the soil surface to retain moisture.

This process verifies that the plant root structure is saturated with water and has minerals and biology available, thereby eliminating any stresses as it settles into its new home. With this kind of treatment, new transplants will relish bright sunlight rather than shriveling up and possibly dying from heat stress.

Amending Perennial Crops

Some simple long-term strategies for fruit trees employ the same set of tools, cover crops, mulches, rock dust mixtures, drenches, and foliar sprays. In the fall apply broad-spectrum foliar sprays to fruit trees before the leaves fall to provide nutrients for next spring's blossoms. Apply foliar sprays to the tree's bark if the leaves have fallen in late fall or early spring. Try making amendments using the fruit of the tree to be amended. Apply additional amendments based on specific mineral needs. Foliar sprays throughout the growing season will significantly improve fruit growth and tree health. A regular spraying interval is important because the plants will rely on the applications as the season goes on. Consider varying the calcium-to-phosphorus ratios to support the phases of plant growth.

Notice the complete absence of spots and discoloration on these peach leaves and fruits. The only sprays applied to this vibrantly healthy tree are my homemade mineral and biological amendments—no pesticides allowed.

Amendment Strategies
Throughout the Growth Cycle

The nutrient needs of plants change during their growth cycle. Germination, vegetative growth, reproduction, and senescence each requires its own unique array of minerals to support the metabolic processes taking place in the plant during that growth phase. Thus each growth phase is a unique opportunity to significantly affect overall plant health. Learn what minerals are needed during these key development phases and the amendments available that will provide them. Foliar sprays and drenches are the best delivery systems to meet the short-term need during a particular phase of growth, such as calcium during fruit ripening to increase fruit firmness. (Details on how to apply foliar sprays and drenches are in chapter 3.)

Procuring Good Seeds

Before focusing on how to optimize plant performance in the various stages of growth, it's important to consider the impact of seed quality on growth. The new garden model recognizes that all seeds are not the same: Some batches of seed are of better quality than others; some are open-pollinated types that will reproduce true to type; some are hybrids; some are genetically modified. A most important recognition is that selecting and saving good seeds is important when the goal is to grow high-quality food. Next year's seeds may be improved if the parent plants are grown in the right mineral and biological soil conditions. George Washington stated, "Bad seed is a robbery of the worst kind: for your pocketbook not only suffers by it, but your preparations are lost and a season passes away unimproved." This quote recognizes the present genetic potential of a seed and the possibility of improving that seed year after year. Of course, in order to reap any harvest at all, planting is a necessity. Select the best-quality seeds available to you and sow them. Plants will not grow if you don't put seeds in the ground.

Seed growers grade seeds on a scale from 1 to 10, and there is no guarantee that seeds in little packages on seed racks at the store are high-grade seeds. Choosing to buy packaged seeds year after year makes it impossible to pursue the annual plant improvements suggested by George Washington.

Saving seeds from the garden is a worthy endeavor, but keep in mind that hybrid varieties are the result of deliberate cross-pollination of two different parent varieties. Seeds produced by hybrid plants may not exhibit the same characteristics in following generations. And saving seed from genetically modified varieties is out of the question. The genetics of these seeds have been altered in laboratories by humans to produce seeds that would not arise in nature. The plants that sprout from such seeds have genetic makeups that are foreign to our bodies. Our digestive systems would never have experienced the particular foods these plants produce, and eating them may result in undesirable reactions.

Purchasing seeds from reputable sources is a better alternative than buying any old seeds available on the seed rack. Saving seeds and trading seeds with other gardeners is better still. Saving seeds is a natural selection process that allows the seed strain to become accustomed to a specific environment. Saving and using these seeds in following years is a most rewarding proposition. In many cases saving seeds is easy and worth doing just to see the results. Saving seeds of a particular variety for several generations will result in a crop that has gradually adjusted to local soil and climate characteristics. Garden time again.

Heirloom seeds are a good place to start. These are the plants of the past whose seeds were saved by growers because they exhibited desirable qualities like good flavor, cold tolerance, bountiful yield, long storage life, and early or late harvest. Local heirloom varieties may already be familiar with the local climate and perhaps soil type. By contrast, commercial plant breeders often have different goals, selecting plants that produce fruits that travel well, are bruise-tolerant, look good on the shelf, and are otherwise suitable for shipping long distances.

Wild seeds are the least adulterated choices and include the seeds of edible weeds that grow locally: stinging nettles, dandelions, purslane, and lamb's-quarter to name a few. These plants are very nutritious, easy to grow, make great cover crops, may be eaten or used to make powerful amendments, and grow like weeds!

The Seedling Stage

A plant starts as a seed and ends as a seed. Seed germination is the most important development phase of a plant, and the one with the most impact on its overall health. The challenge for gardeners and farmers is to eliminate any stresses that a plant may encounter throughout its life so that it may reach its genetic potential and produce seeds of better quality than the seed that started the season. Just as the nutrition and environmental health of the first two years of a child's life define many of her developmental capacities, so do the first weeks of a plant's life. Initial seed development from sprouting to framing defines a plant's overall development capacity.

The quality of the environment and available nutrition at the time of seed germination determines the number, size, and quality of the fruit the plant will eventually produce. Cell division and cell growth that occur during germination require adequate proportions of calcium, potassium, boron, and manganese, and without success at this early stage, a plant cannot produce high-quality fruits. Calcium may be found in limestone quarry rock dusts; fermented plant juice of nettle, dandelion, or sassafras leaves; vinegar extractions of eggshells, oyster shells, or cow bones; or leaf-mold-fermented carrot tops. Potassium is found in these amendments as well in slightly different proportions. Boron can be found in vinegar extractions of oyster shells and cow bones and the fermented plant juice of stinging nettles, chickweed, and leaf-mold-fermented carrot tops. Manganese may be found in some basalt quarry rock dusts and in fermented plant juice of quack grass, dandelion, stinging nettle, or sassafras leaves. Use the information in appendix C and E as a guide for selecting plants with desired mineral contents.

In order to be sure that the seed gets all the nutrition required from the start, soak the seeds for 10 to 15 minutes in a solution of mineral-enriched good water. Stinging-nettle- or dandelion-fermented plant juice provides a broad spectrum of minerals and other beneficial compounds all in plant-available forms. Soaking the seeds in a leaf-mold-fermented amendment of the same plant would provide them with minerals and other beneficial compounds unique to that plant. Why guess what a plant seed needs? Simply give the seed its own riches. It's also helpful to add a biological amendment (IMO #2 or #4, leaf mold biology, lactic acid bacteria, or raw milk) to the soil before sowing seeds.

Large seeds such as beet seeds are relatively easy to soak and then remove from the soaking jar to plant. Small seeds may be more difficult. Transferring individual wet seeds into small pots for indoor growing is relatively straightforward, but when direct-seeding very small seeds such as carrot seeds outdoors, it's best to first cast the unsoaked seeds onto the soil and then water the bed with the mineral amendment solution. Repeated applications of water-soluble amendments will ensure that minerals are available during germination.

After sowing, continue applying mineral-enriched good water to the seeds and sprouts, especially during the first few weeks of growth. Remember that these plant-based amendments contain not only minerals in plant-available forms, but also a broad spectrum of other compounds that will nurture the plants.

Biology may also be added to the soil a week or two prior to planting or transplanting in order to establish a rich biological environment for roots to interact with as soon as they enter the ground. Leaf mold biology provides the most diverse selection of local biology, but lactic acid bacteria or raw milk may be used as well. Once roots have become established and leaves unfurl, the plant will use some of the energy created during photosynthesis to selectively feed the soil biology, turning the ecosystem into its own digestive system. It is exciting to watch the young plants jump out of the ground with a vigor that unmistakably signals their good health. Acknowledging that this quality of plant development is different from past experience is empowering. Experiment with various recipes and timing of application. Keep records of what you try, and compare results.

Keep in mind that every crop has an optimal soil temperature for germination and growth. Making sure these temperatures are met gives the plant the best opportunity to thrive. Using a heat mat is the easiest way to attain optimal soil temperatures when starting seeds inside. A feedback system allows the soil to be kept at that optimal temperature night and day. Outdoors, use a thermometer to monitor soil temperature, and wait for the right temperature range before you plant.

The Vegetative Stage

For vegetative crops like lettuce, kale, and spinach, photosynthetic efficiency is a limiting factor to growth. Magnesium, iron, manganese, and phosphorus are needed to maximize photosynthetic efficiency. Once seedlings have begun to photosynthesize, apply these minerals as foliar

This leek flower cluster will be covered with a wide variety of pollinators once the hundreds of small individual blossoms open.

sprays. Sources of magnesium include vinegar-extracted cow bones, fermented plant juice of stinging nettle or dandelion, and leaf-mold-fermented dandelion or carrot tops. Iron, although present in the soil, is often not in a form that plants can utilize. Leaf-mold-fermented carrot tops and fermented plant juice of dandelion, quack grass, and fruits contain iron in forms that the plant can use. As noted above, sources of manganese include some basalt quarry rock dusts and fermented plant juice of quack grass, dandelion, stinging nettle, or sassafras leaves.

Phosphorus is available in vinegar-extracted cow bones, fermented plant juice of fruits and leaf-mold-fermented carrot tops. Again, all these amendments also contain other minerals and compounds that will contribute to plant health.

During vegetative growth observe leaves for indications of mineral deficiencies. Refer to "Plant Mineral Deficiency Indicators," appendix D, for a helpful summary. As soon as you notice a deficiency, apply a foliar spray or drench plants and soil to address the problem. Biological amendments may be used to stimulate the ecosystem. Since most of the amendment recipes described in this book have plant-based broad-spectrum mineral proportions, the risks of toxicity are lower than if using commercially produced single-element mineral compounds, especially in the case of trace minerals needed in concentrations as small as single-digit parts per million.

The Reproductive Stage

During the reproductive phase, plant nutrition priorities change. Mineral and nutrient needs of the flowers and fruit become primary, and minerals and nutrients are sometimes relocated from the leaves and stalks. Carbohydrates produced via photosynthesis are also prioritized to the flowers and fruit. Phosphorous-rich foliar sprays such as vinegar-extracted bones may be used to facilitate the flowering phase and calcium-rich foliar sprays such as vinegar-extracted oyster or egg shells may be used to facilitate the fruiting phase. Each may be used in conjunction with a broad-spectrum mineral amendment like dandelion- or nettle-fermented plant juice.

The stage just before a plant drops its leaves is an opportunity to provide nutrients for next year's life cycle. One example cited by plant nutrition expert John Kempf is a foliar spray of cobalt, which may delay senescence and thereby increase crop yield. Local sources of cobalt include fermented plant juice of peach, apple, and mugwort. A vinegar extraction of the fermented plant juice residue may yield cobalt in a higher concentration.

Summing Up

One finds that there is always enough in the garden. Learn to use the new plant model to form ideas about what you want to accomplish and then look around your property and your neighborhood to see what materials are available to help achieve these goals. That pile of wood chips found after tree damage from a storm could be a layer on your compost pile or a mulch for a garden path. Those oyster shells that the local restaurant throws away could make an excellent vinegar-extracted mineral source. Leaves piled beside roadways in the fall could mulch a garlic bed, be layered onto the compost pile, or just be stored for future use. That field of dandelions or stand of mugwort could be turned into a large batch of fermented plant juice. Learn to identify resources that are local and free and use them to complete the larger cycles of life, death, and decay. These are the principles of sustainable, regenerative gardening.

As time goes by and your landscape begins to respond to your efforts, pause on occasion to enjoy observing the diversity and numbers of pollinators that increase. Notice the weeds change year after year as the soil mineral composition changes. You may even notice an increase in the population of some pests, but these will be balanced by an increase in their natural predators as the entire ecosystem adapts to the stimulation of local diversity. Notice changes in the flavors of your crops and improvement in their size and their ability to keep well in storage. Relish the satisfaction to be had by stimulating nature's processes rather than trying to outwit its flow by using chemicals.

CHAPTER 3

Sustainable, Regenerative Garden Tools

I n the old garden model, rototillers, shovels, wheelbarrows, and hoses are the essential tools that come to mind. But in the new garden model, the emphasis is on tools that directly affect the soil ecosystem and plant health, including water, mineral and biological amendments, compost, cover crops, mulches, and weeds. These are the tools gardeners use to balance mineral proportions within the soil, introduce diverse biology, increase the humus content and exchange capacity, improve the flow of energy, and maintain appropriate hydration within the soil and plant systems.

All water is not equal, and this chapter begins with a discussion of what constitutes "good water" and its importance in the making and application of mineral and biological amendments in the form of foliar sprays and drenches. The compost pile is an amazing digestive system in which local biology decomposes lignin, cellulose, chitin, lipids, and minerals into rich humus that can then be used to regenerate the garden soil. Cover crops protect the soil ecosystem from death by desiccation and also feed the soil biology with sugars they manufacture through photosynthesis. Mulches provide shelter, preventing the evaporation of water needed for the soil biology to survive, as well as supplying abundant organic carbon to feed the soil biology. Weeds can be used as cover crops, mulch, layers on the compost pile, raw materials for making amendments, or nutritious food for the gardener to eat.

Good Water

Water is perhaps the most amazing compound on Earth. Able to dissolve most substances if given sufficient time, water is a carrier of geological information, a medium of communication between biological systems, and a vital component of every living cell. Another fundamentally important characteristic of water is its ability to be structured. The infinite structural characteristics of water are well known to those who live in regions with snowy winters. Every snowflake has a unique pattern. Victor Schauberger, Rudolf Steiner, Theodor Schwenk, Andreas Schulz, and Fritz-Albert Popp are some of the names associated with the concepts about *structured water*. The depths of our understanding of water is changing as books such as *Cells, Gels and the Engines of Life* and *The Fourth Phase of Water* by Gerald H. Pollack are redefining the known phases of water and the mechanisms by which minerals move in and out of cells. Another important characteristic of water is the concept of hardness—the measure of carbonates, bicarbonates, and/or sulfates in the water. Hard water not only clogs irrigation systems with the deposit and buildup of salts but will significantly reduce the effectiveness of the amendments provided to plants. Finally, the pollution in a given water source needs to be considered. A lifetime spent studying water would not be boring.

Spring water, well water, and water from a stream or pond may be acceptable for watering the garden. Whatever the source, be sure to test the water for pollutants and hardness prior to use if no testing has been done previously to be sure that the water used for irrigation is without harmful chemicals and has a low measure of hardness. According to John Kempf, even small amounts of hardness, 150 parts per million, will reduce amendment effectiveness by 70 percent. Water with hardness values of less than 70 parts per million is desirable. Filtering may be used to remove chemical constituents, but the hardness of any water can be improved only via reverse osmosis, which has its own set of concerns. There are costs, volume restrictions, and the wastewater must be dealt with.

Municipal water is usually treated with chlorine, fluorine, and other chemicals. It represents the bad end of the good water spectrum. Chlorine, added for the purposes of killing waterborne bacteria, will negatively affect the soil biology. You can improve municipality water by filtering it or, as a minimum, leaving it to breathe in a bucket for a few hours before use to allow some undesirable constituents to outgas.

Rainwater

Rainwater is the best water to use for watering plants and for making the amendments described in this book. This is "good water." Drops of rainwater originate in the relatively clean environment high in the sky. With an electrical conductivity near zero, rainwater contains little or no unwanted carbonate or bicarbonate ions. Rainwater is free, and it is easy to gather. Those lucky enough to live in an area with adequate rainfall can establish a rainwater collection program.

Collecting rainwater is sustainable and rewarding in many ways. Having a barrel full of rainwater just feels good. A most pleasurable gardening treat is to be in the dry, lush environment of my hoop house while a heavy rain outdoors is filling the rain barrel. The force of gravity pushes the water through a hose into the hoop house as I water the plants inside.

The first step in setting up this type of system is finding a barrel. Old wine barrels are functional and have a pleasant aesthetic. They are contained: The top is sealed so insects, leaves, and other debris are prevented from getting inside. These may be purchased ready for use, with a spigot and drain in the bottom and a port of entry in the top. The barrel needs to breathe, though. As water enters the barrel, air must have a way to escape. A small hole, about ¼ inch (6 mm) in diameter, drilled into the top of the barrel will do. Adjust as necessary. Think "I need a rain barrel." Say it aloud. Then use what comes your way.

Another option is to collect rainwater in a large bucket or can. In this case, make a cover (perhaps a piece of wooden board) to keep out debris and mosquitoes or other insects. Whatever the type of rainwater collection container, it's a good idea to set it upon a sturdy base, allowing enough space to place a bucket or watering can underneath the spigot. The base can be as simple as stacked rocks.

I do not like the look of gutters on my house, or the work required to maintain them, but I have installed a gutter on the back side of a shed. This makes for a highly functional, easy-to-maintain, hidden water collection system. On a sloping site, set up the rainwater collection system as far upslope from the garden space as possible and utilize gravity as the delivery force.

When winter approaches, it is important to empty the barrel before it freezes. Once frozen it will not likely thaw. At 8 pounds per gallon, a 40-gallon (150 L) barrel of frozen water can easily exceed 300 pounds (136 kg), making it difficult or impossible to move. As the temperatures continue to drop, the ice will continue to expand and eventually split the barrel. The barrel may be emptied and left outside for the winter. Turn it upside down or on its side to prevent water from collecting and freezing in it, or move it into a shed. I leave the spigot open on my wine cask rain barrel for the winter so that water may be collected if we get rain. I can collect it and quickly water my hoop house with it before the temperatures drop below freezing. This will work for most containers depending on how much water remains on the bottom to freeze when it does get cold.

It is possible to gather rainwater even during the below-freezing temperatures of winter for watering plants growing in a greenhouse or hoop house. Whenever it does rain, set out buckets to fill. When the buckets freeze solid, bring them into the hoop house and turn them upside down. On a sunny day, ease the bucket off the solid ice tower. Place the solid ice column next to a perennial plant to slowly melt, watering the local vicinity. Snow can be shoveled into buckets, brought into the hoop house to melt, and then distributed as needed. These buckets of frozen water or snow may also be brought into a basement or other warm area to thaw and used for watering indoor plants.

If rainwater collection is not possible, select the best water available with an understanding of its benefits and drawbacks. It's wise to have the water analyzed to establish the amount of pollutants and the degree of hardness of the water source. The information provided here is meant to be a starting point from which to learn how to best respond to the quandaries of not-so-good water when the goal is to grow high-quality, nutrient-rich food. Research and learn what can be done to improve the present situation. Remember that life is tenacious and plants will grow with most available water.

Mixing Amendments into Water

Because some trace minerals are required only in very small concentrations (and excess can be harmful), the challenge of distributing very small amounts of a mineral over a large area comes into play when mixing amendments. In order to be sure that the scant amounts of a mineral are evenly distributed within a mixture, a thorough mixing scheme is essential. It is also worthwhile to add energy and intent to the water delivered to soil and plants. We do not fully understand the complexities of water; nor do we understand the full extent of our capacity to influence water, plants, and the world around us.

Spinning the contents of a bucket of water such that a funnel or vortex is created at the center of the container produces excellent mixture of the solutions in the bucket and also adds energy to the solution. Stir the water in one direction for a period of time, and then reverse the direction and stir the water the other way around for a similar period of time. Alternate the direction of stirring at least seven times to assure proper mixing. The vortex that is created in this manner causes the water molecules to move at different rates of speed, depending on how far from the center of the bucket a molecule may be. Those near the outside edge of the bucket must travel a longer distance than those near the vortex, and thus they move faster. This difference in speed causes a shearing effect between adjacent water molecules, resulting in good mixing, and these shear forces add heat energy.

The mineral amendment recipes in this book contain minerals in ionic forms—they are charged particles. To add additional energy to the

My mixing bucket for amendments has magnets equally spaced around the outside perimeter, alternating polarity from north to south to north, and so on. The magnetic field adds energy to the mineral ions in the water solution.

water solution when I mix amendments, I position powerful rare earth magnets on the outside perimeter of a bucket. When a charged particle moves through a magnetic field, that field adds energy to the particle. This added energy, in conjunction with the energy associated with the shearing action of mixing, changes the molecular structure of the water, thus the term *structured water*. The literature on structured water states that the water becomes more stable, surface tension is reduced, and viscosity is increased; these effects may last for some time. Fascinating! This is my surfactant, the reduction of surface tension so that the water sprayed onto leaves spreads on the surface rather than balling up.

In the world of energies and their effects, here is a final thought to consider. The thoughts that course through one's mind can also affect the energy structure of the mixture in a bucket as one stirs. Intent defines the future. You may find this hard to imagine, but I encourage you to try it out. As you stir, think and state your intent of nurturing the plants, of the great flavors to be tasted, of the insects that will enjoy pollinating the plants, or any other ideas that come to mind. You may begin to experience a feeling of acceptance, an introduction to the great expanse that is the power of the earth, a connection with the plants and animals within the soil, an ethereal feeling of belonging to something that is more vital and purposeful than the everyday world of shopping centers, television, or texting. When you stir water in a bucket for several minutes in one direction, then switch directions for six more repetitions, that's an investment of real time and energy. Rather than feeling miffed or frustrated about this seemingly silly step of the process, let it be an opportunity to look around at the environment and take note of what is going on. Look closely, and more closely still, until something previously unnoticed comes to light. Smile and appreciate the clean air, the puffy clouds, the sunlight reflecting off the leaves of the trees, and the breath moving through the body. Know that the water you are preparing for the garden is rich with plant-available nutrition and an energy generated by you.

Foliar Sprays and Drenches

Foliar spraying and drenching are two effective ways to supply nutrients, biology, and water to plants. Foliar spraying is simply applying a mist onto plant leaves and stems. The most efficient absorption is through the leaf surface, but if a plant is leafless, you can mist the surface of the bark, which will absorb some of the foliar spray. The water and nutrients of the spray solution are absorbed through the plant surfaces and transported throughout the plant via the phloem pathway. Foliar spraying is the fastest way to provide nutrition to a plant. The leaves can absorb and distribute the spray solution in hours or even minutes, especially under the right conditions. In contrast, fertilizers applied to the soil in the form of a powder or rock dust need to be digested by the soil ecosystem before roots can draw them into the plant. This digestion process can take weeks or longer. This is why applying rock dusts is best done in late fall through early spring (long-term mineralization).

You can measure plant response to foliar spraying by applying the foliar spray to a portion of a crop and then measuring the refractive index of the sprayed and unsprayed plants and comparing them. (See chapter 4 for instructions on how to measure refractive index.) For example, make comparisons a couple of hours after spraying and then again the following day. It is always wise to apply amendments on a small experimental area first to be sure the plants like it. Once you become familiar with an amendment application, then it may be used regularly and more extensively. Be sure to record such data in your garden notebook!

Deciding What to Apply

Be thoughtful about your choice of amendments when you prepare a foliar spray solution. This may be as simple as selecting an amendment that has a broad spectrum or as complex as doing the math to figure out the mineral proportions to accommodate a specific growth phase or nuance of a specific plant. Individual crops may have very different mineral needs throughout their growing phases. Soil test information may provide insights about deficiencies to address with a particular foliar application.

Broad-spectrum mineral amendments, like fermented plant juice of stinging nettle, dandelion, or leaf-mold-fermented carrot tops, are great starting points when preparing a foliar spray because they contain quantities of the macro- and microminerals wanted and in proportions that most plants will like as well as other complex compounds. In order to accommodate a specific growth phase of a plant, the proportion of calcium may be increased by adding vinegar-extracted oyster shells or eggshells. The proportion of phosphorus and calcium may be increased by adding vinegar extractions of cow bones. For information on choosing plants to make customized amendments, see "Weeds and Crop Plants" on page 113 and the mineral composition of many amendments in appendix E.

Another approach is to use an amendment made by fermenting the same type of plant as the plants to be amended. One example is the yearly tomato crop: Make an amendment using damaged, insect-eaten, or not-quite-ripe tomatoes in the fall. This can be done using either leaf mold fermentation or sugar fermentation (fermented plant juice). Use this amendment to foliar spray the tomato crop the following year. This strategy assures that the plants will receive minerals in appropriate proportions without your needing even to think about sourcing or mixing them.

It is best practice to maintain plant health so that plants do not succumb to pathogens or insect pressure, but occasionally a problem may arise. Foliar sprays that contain biological amendments can be applied to combat an existing infection (such as powdery mildew) caused by airborne pathogens. The pathogens are generally always present in the local environment, but they will not infect plants until the plant leaf surface is compromised in some way. Stresses such as lack of water or poor nutrition can deplete the natural lipid layer on the surface of leaves, allowing pathogens to take hold by penetrating the degraded cuticle layer and infecting the tissue within the leaf. When powdery mildew affects your squash plant in late summer, it is because the plant is not getting the nutrition needed to maintain this lipid layer and the mildew takes advantage of the unprotected leaf. Biological amendments, in conjunction with mineral amendments, will increase the biological competition on the leaf surface and provide nutrition needed to restore or improve plant health. The pathogen must then compete with the added biology on the leaf for food, and the minerals will help increase overall plant health. I have experimented with this approach a few times but have not done a good job of monitoring and documenting results. I offer this as another area for experimentation.

Setting a Schedule

As mentioned in the previous chapter, it's good to establish a regular schedule for foliar application that the plant can depend on throughout the growing season. Consistency is important, not only for steady plant growth but also to maintain plant health and thus prevent the plants from coming under attack by pathogens or pests. Since most soils exhibit one or more mineral deficiencies that will take time to correct, a short-term foliar spray program feeds the plant what the soil cannot. The plant learns to rely on a steady supply of these foliar-sprayed nutrients. The consequence of not getting those nutrients is poorer health and potential damage from pests and disease. One year, because I was busy, I stopped my regular foliar spray regimen in the middle of the growing season. I observed that the rate of plant development changed: Fig fruits stopped growing altogether, and tomato fruits significantly slowed development.

As mentioned in chapter 1, the sugar content of plant sap is an indicator of overall plant health. When sugar content is high, insect pests are unable to successfully attack the plant, because they do not have the capacity to produce the digestive enzymes needed to break down sugars above a threshold level. Plants with sap sugar content above this threshold are not a food source for that insect. As available nutrients become depleted, whatever the reason, sugar content of plant sap also declines. Once the sugar content of plant sap drops below the threshold, insect pests move in and do what they are supposed to do: Eat those lower-quality plants. So when a single potato beetle is spotted in the potato patch, take action. Its presence indicates the plants need attention! If a crop is under attack, be sure the soil has adequate water and that the soil ecosystem is thriving. Next, apply a broad-spectrum mineral amendment as a foliar spray, which may quickly and effectively resolve the depleted nutrient issue. More than one application may be needed, but once the plants have taken in the nutrition required to increase their sap sugar content, enjoy the satisfaction of seeing the beetles disappear. Developing good observation skills and acting as soon as problems are identified is a way of becoming in tune with the garden ecosystem.

In general, it's best to apply foliar sprays early in the morning before the sun comes up, while there is still dew on the leaves. Foliar spraying after the sun has set will work as well. It is desirable to have the foliar spray last as long as possible on the leaf surface, to be absorbed into the phloem pathway and distributed into the plant. When the sun is in the sky, the leaf surfaces are hot and any liquid on them will quickly

A simple mister-type spray bottle or a manual pressurized pump spray bottle works well for applying foliar sprays in home gardens.

evaporate. Leaf stomata are fully open and the plant transpiration rate is high. The flow of moisture from the soil to the leaves through the xylem pathway is at a maximum. In the very early morning hours, however, the stomata are closed and the xylem flow is at a minimum. Nutrients absorbed into the phloem pathway will have more time to mix with the slow-moving xylem flow, and thus will be better distributed throughout the plant sinks. Conditions of high humidity increase absorption, too, but spraying just before a rainfall is not a good idea, because the rainwater may wash the spray solution off the leaves before it can be absorbed. In the realities of day-to-day life, however, there may be only a small window to complete the task. Stick to a regular schedule as best you can.

Schedules for foliar spraying may vary depending on many factors. One possible schedule is to simply foliar spray all crops every 7 to 10 days using a broad-spectrum mineral amendment. For fruit trees, a basic guideline is to apply a foliar spray once in the fall, again in early spring, and another time during fruit-set. Sometimes I choose to spray on a windy day, knowing that everything downwind of the sprayed area will also benefit from the application.

Making and Applying a Foliar Spray

I use an electrical conductivity meter when mixing foliar sprays. The electrical conductivity of rainwater is close to zero, but the mineral amendments recipes I use are loaded with ions that increase the conductivity of the solution considerably. The recommended dilution rates for these amendments are 1:500 (amendment:water) or 1:1000. Even at these dilution rates, the electrical conductivity of the solution will increase with each additional type of amendment added. Conductivity is measured in millisiemens per centimeter (mS/cm). I monitor the electrical conductivity as amendments are added, and in my experience solutions that have values between 1.5 and 2.5 mS/cm do not seem to burn plant tissue.

Foliar sprays have a tendency to ball up on the plant surface because of the surface tension of water, but it is desirable for the spray droplets to spread out on the leaf for better absorption. One way to overcome surface tension is to add a wetting agent to the spray. I've found, though, that foliar sprays made with homemade amendments and rainwater do not have this tendency to bead up. This may also result from the particular way in which I mix the spray solution in a bucket with magnets mounted on it, as described earlier in this chapter. The surface tension of water is reduced when exposed to magnetic fields in this manner—the structural change of the water molecules acts as a kind of wetting agent. The water also becomes more stable and accelerates plant growth. It is interesting to read research papers on this subject, and I've included some in the bibliography.

To apply a foliar spray to a large area, a backpack pump spray system like this one may be more efficient than a smaller spray bottle system.

There's a wide range of devices for applying foliar sprays, from simple hand-pump spray bottles to gas-powered backpack sprayers. The area to be sprayed, the frequency of application, and the funds available will determine which device is best. The spray bottle under the kitchen sink may be the best tool to start with. The most effective spray technique is to apply a light mist to the plant surface. Avoid spraying so intensively that the water ends up running off the leaves.

Making and Applying a Soil Drench

A drench is simply the application of a diluted water-based amendment directly to the soil using a watering can or irrigation system. The amendment immediately mixes with the soil solution and may be absorbed by the plant roots and into the xylem pathway. There are particular situations where applying a drench is a good strategy. For example, if a mineral is not very mobile in the phloem pathway, as is the case for calcium and manganese, applying it via a drench may be more effective than a foliar application. Also, if a soil's exchange capacity is low, then periodic drenching will help maintain a supply of nutrients within the soil solution, preventing a temporary deficiency of nutrients that could otherwise cause plant health to deteriorate.

I often apply a drench shortly after a rain. The abundance of water in the soil at such times helps to distribute the minerals in the drench. As with foliar applications, it's advisable to establish a regular application schedule for soil drenches to provide a steady nutrition program that plants can depend on throughout the growing season.

Compost

Making and using compost is a cyclic process that plays itself out in garden time. How long it takes for a compost pile to decompose determines how often a new one needs to be made. How much is used each year determines how big the pile has to be. Digestion of the wide variety of organic materials that are tossed onto a compost pile requires a diverse biology: Bacteria to digest the chitins and cellulose, fungi to digest lignins, and as for archaea, who knows precisely what function they play in this process? Compost piles can run hot when they are turned often, which promotes aerobic bacterial function to drive decomposition.

Or they can be left undisturbed, and over a longer period of time the fungal community will act as well, promoting humification.

Location is an important consideration. A pile could be considered unsightly and thus be relegated to a corner of the yard far away from the garden. But once it's time to move the finished compost to the garden, that choice of location may be a cause for regret. Another possibility is to set up the compost pile in the garden itself or in a spot that will become a garden space in the future. In that scenario, once the compost has been used up, the soil beneath is usually more fertile than it was before the pile was constructed there. In this way a compost pile becomes part of the overall regenerative plan.

Table 3.1. Compost Materials

Material	Function	Percentage of Pile (by volume)	Sources
Green materials	Supply protein (nitrogen)	50	Weeds, kitchen scraps, grass, manure, fish
Brown materials	Supply carbon	30–50	Shredded leaves, hay, straw, sawdust
Clay and humus	Storage of cations, anions, and humus; housing	10	Mature compost, soil, silts, clay, carbon
Minerals	Improve mineral balance of soil	5	Rock dusts, vinegar extracts, ocean products, fermentations, IMO #4, raw milk
Biology	Improve biological diversity	Add as available	Water extractions, leaf mold biology, lactic acid bacteria, raw milks, small sticks, IMO #4
Air and water	Facilitate overall function of pile	Add to pile as needed	Small sticks (to provide air spaces), rain (water)

A compost pile may be thought of as a lasagna, with layers repeated many times as materials become available. A simple model may define the layers as browns and greens. Brown layers include dried crushed leaves, straw, hay, wood chips, and other organic carbon-rich materials. Green layers include kitchen scraps, fresh manure, and fresh grass clippings or other freshly cut plant matter. (See "Mulches" on page 89 for more details about these types of materials.) A more sophisticated compost model recognizes a few other layers to incorporate into the pile: minerals in the form of rock dusts; weeds, especially those rich in mineral content; local biology sources, including some of the recipes in this book; soil, including IMO #4; and twigs, water, and compost itself. These additional layers will provide housing and sustenance for the biology needed to decompose the pile and minerals to help balance the soil that the finished compost will be applied to. Adding these materials to the compost pile thus becomes part of a long-term strategy for amending the soil.

The base layer of a compost pile should be slender twigs less than ⅜ inch (1 cm) in diameter. These may be green or brown. Fungi that live under the bark of these green twigs will propagate in the pile and help to digest lignins. Using twigs as a base also allows air to penetrate the pile, which is important to facilitate aerobic digestion. Next, add alternating brown and green layers, each about 1 inch (2.5 cm) thick. After each duo of brown and green layers, spread a layer of finished compost or soil also about 1 inch thick. You can use garden soil, imported soil, or a combination of these. This layer will provide habitat for the biology needed to digest the entire pile. Add a biological amendment, too, such as leaf mold biology, lactic acid bacteria, raw milk, IMO #2, or IMO #4. The pile environment will provide excellent food and housing, and the biology will thrive.

Additional layers of rock dust or crushed shells to supply minerals could be added, which will be digested over time by the biology into plant-available forms. Refer to soil test results to determine which minerals and what amounts to add to a compost pile, or strive for the ideal soil mineral proportions listed in appendix B to make a minerally balanced compost product.

Continue adding materials as they become available and as time permits. My brother-in-law likes to fish in the ocean. He uses mackerel as bait, and he gives any leftover mackerel to me. This is a wonderfully oily fish and perfect for the compost pile. I add them onto the pile as follows: A layer of hay, then IMO #4, water, a layer of whole fish, more straw, more IMO #4, and more water. There are no resulting fishy smells that would attract animals.

Be sure to keep the pile from drying out. Depending on the frequency and amount of rain in your local area, you may need to add some water to the pile, or none at all. Top off the pile with a brown layer to keep in the existing moisture but keep out any additional water. Let the pile digest undisturbed for the next 12 months. During that time it will transform into the finest dark, crumbly, fresh-smelling finished compost.

Late fall or early winter is a good time to start a compost pile; add materials to it for about a year and then stop. I maintain three compost piles at all times: the one I am in the process of adding to, the one that is digesting material for use the following year, and the finished pile that I take material from.

Cover Crops

I strive to keep my garden covered at all times with a cover crop or a mulch. Growing cover crops helps meet the goal of having plants growing in the soil at all times, feeding the soil ecosystem with the sugars the crops create during photosynthesis. My favorite cover crop is the weeds that have planted themselves naturally in a bed. I intervene by choosing which weeds I allow to go to seed, so there is some degree of selectivity in my cover cropping. I also propagate certain plants as cover crops, including garlic, dill, lettuces, cilantro, parsnips, collards, and kale, as well as dandelion, dock, stinging nettles, purslane, red clover, and sorrels. I allow these plants to form seeds, and in late fall I grab handfuls of the seeds right off the stalks and cast them around the garden space and any other places I wish to see them grow. This way, I know that when spring arrives, something is sure to sprout in any unmulched patches of soil, and that those sprouts usually will be something I can eat. Also, when the soil is disturbed to plant a crop in these locations, the cover crop seeds often germinate as well, resulting in a mixed bed of crops. There's always the option of removing the cover crop if I want to plant something else in the space instead or if there is excessive competition.

My garden soil is loose owing to years of this style of management, and so in the early spring, it's easy for me to remove whatever cover crop has sprouted if I choose. But if I don't need the space for planting a food crop, I allow the cover crop to continue to grow. I know the soil will be covered and nurtured by these lovely plants until the time when I'm ready to use that part of the garden for another purpose. Since I can eat most of the plants I choose for cover crops, I can harvest them as the growing season unfolds, as well. Those early garlic bulbs have a lovely mellow flavor, the young dill is a treat, the red clover flowers make a wonderful tea, and the dandelion and nettle leaves are nutritious and tasty. If I leave a cover crop in place for a full season, I may periodically knock down the plant tops with a scythe or string trimmer to control seed production, to add a green manure to the soil surface, and to allow those seeds underneath a chance to grow. I do not disturb the cover crop roots unless I plan to replant the space with a new crop, because these roots still have value in the soil. They become food for the soil ecosystem, or they regrow. The deep taproots of plants like dandelions, dock, and parsnips decay, leaving spaces that serve as passages for worms and other root systems to come. By taking advantage of garden time, doing "nothing" is revealed as a wise cover cropping strategy.

Whenever a vegetable crop is harvested, plant another crop as soon as possible. Sometimes this may be a cover crop. When feasible, it is sensible to sow the seeds of the next crop prior to harvesting. This strategy keeps those seeds moist and hidden from the birds or other plunderers. The seed will germinate readily in the shade beneath the existing crop. When the harvested crop is removed, the seeds have already germinated and are on their way. An example of this is to cast oat and pea seeds around a bed of garlic a couple of weeks prior to harvest. The oats and peas will be off to a good start by the time the garlic is harvested. Soil exposure to drying air and sun is reduced, and the soil ecosystem has a chance to transition to the desires of the newly planted crops.

On those occasions when you are purposefully planting a cover crop in a vacant bed, it is best to sow a variety of species. The new garden model recognizes the strength of diversity. Experiment with combinations of grains, herbs, and legumes, as many varieties as possible. A wider range of plant types can support a greater variety of soil biology. One option is to simply allow the plants that want to grow there to prosper. Another is to plant excess seeds not planted in previous years. I purposely grow extra bean and pea plants specifically so I can harvest and dry their seeds to sow in future years as cover crops.

When you're planting cover crops, it is worth noting which types will have the capacity to winter over and which will winter kill. This will depend on your local climate—the length and harshness of winters in your area—but if the cover crop will winter over, be sure to have a strategy for dealing with it the following spring, and be sure to follow through with that strategy. Otherwise your cover crop can become a future problem. Allowing winter rye to winter over, for example, and then failing to cut it down before it goes to seed in early summer can lead to more work in the future trying to get rid of the self-sown winter rye.

Mulches

Mulching the soil surface protects the soil ecosystem from drying out and prevents weeds from growing. It serves as an organic carbon source for the soil biology and in the end, becomes humus that enriches the soil. Mulch has many purposes. It's used to define a path through the garden and to suppress unwanted plants in a flower bed. It can also be used as a deliberate food source for soil biology or to start a new garden. A ring of bark mulch or crushed shells around the trunk of fruit trees, berries, or shrubs provides food for the soil biology, suppresses weeds around the trunk, and helps avoid the incidence of string trimmer damage to the bark.

Using mulch for multiple purposes contributes to a regenerative soil enrichment program. Mulch materials include bark, sawdust, wood chips, crushed leaves, straw, hay, grass, weeds, and crushed stone, bones, or shells. Each of these has its own best purposes, mainly determined by the time it takes for the material to break down and become part of the soil itself.

Crushed leaves are a staple in any garden. Leaves that are not chopped up persist longer, but leaves that have been run through the lawn mower or a shredder will break down quickly, in garden time, providing short-term soil coverage and an excellent source of organic carbon in the near term. For any areas that are not covered by a cover crop at the end of the growing season, spread a layer of crushed leaves a few inches thick. This will protect the soil until spring, provide some early weed suppression, and break down into humus relatively quickly. Biological and mineral amendments may be added to the crushed leaves or any mulch at any time to accelerate decomposition and to add minerals that are lacking in your soil.

When starting a garden from scratch, lightweight mulches such as hay, straw, or crushed leaves will persist long enough to kill the underlying grass, but they will also break down in a timely manner to support planting once the sod has decomposed.

An example of the application of a mulch in a regenerative garden is the use of crushed leaves when planting garlic. The planting bed is prepared by loosening the soil 4 to 6 inches (10–15 cm) deep and removing all rocks. Compost, rock dusts, and IMO #4 can be added, and the soil is rubbed by hand to eliminate any large clumps. The garlic cloves are planted and watered with a solution that contains biological and mineral amendments. The final step is to apply a layer of crushed leaves 6 to 8 inches (15–20 cm) deep over the entire surface of the bed. This is watered well to provide needed moisture and to mat down the leaves so that the wind will not blow them away. In southern New England planting time is in the fall, generally around the middle of October. The leaf mulch not only protects the soil from drying out and prevents weed seed germination, but also keeps the soil warm well into the winter months, allowing the garlic to establish roots before going dormant for the winter.

When spring arrives and the emerging garlic shoots are working to push through the mulch, simply move the leaves aside just enough to allow the shoots to poke through. This will maintain the mulch cover, retain moisture, and continue to prevent most unwanted plants from growing, eliminating the need to weed the bed. As the season continues, weeding needs should be minimal. After the garlic is pulled, a different crop can be planted, or just let the leaves continue to decay and protect the soil ecosystem. One option is to plant winter squash near the garlic bed. After the garlic is harvested in mid-July, the squash will have significant real estate to spread into, and the squash foliage will serve as a living mulch that also prevents drying out of the soil. If time and resources permit, add mineral rock dusts and/or other biological amendments after the garlic harvest to improve mineral proportions or promote biological diversity. Rock dust or crushed shells may be spread on the surface. A drench containing minerals and biology may be applied to the surface using a watering can or irrigation system. The following spring, the soil in this area will be substantially improved by this process.

Weeds

Weeds have a bad name. But although the term is used with disdain, it doesn't take much research to find that weeds are some of the most nutritious foods we have. They accumulate minerals that can be extracted to produce, arguably, the best plant amendments available, and they will grow in almost any conditions, making them essential cover crops or mulches. And they grow quickly, like weeds! Learning to use these plants that want to grow locally rather than fighting to eliminate them can provide a cover crop, a source of mulch, materials for making amendments, a food source for the soil biology, and harvest for the table.

Weeds grow in soils that are minerally deficient, bringing the minerals from within the soil up to the surface. The proportions of minerals within the plants may differ from the mineral proportions in the topsoil. As weeds grow and die each year, the spent plant tops decompose, which changes the soil mineral proportions over time. In turn, the change in the soil mineral proportions will make the growing environment favorable for different species of weeds. Thus, the weeds present in a garden space are telling a story about the minerals that are missing within that soil. Two marvelous books—*Weeds and Why They Grow* and *When Weeds Talk* by Jay L. McCaman—provide information about weeds as indicators of soil conditions and can serve as guides to understanding the soil mineral content of an area just by identifying and learning about the weeds growing there.

Stinging Nettle

Stinging nettle (*Urtica dioica*) is the multivitamin of the plant world. It grows all over the world in rich soils. Take a look near mineral-rich piles of manure and you'll probably find a patch of nettle. A person could survive eating nothing more than stinging nettles for some period of time. I have a friend whose mother survived the journey out of the ghetto in Poland during World War II by relying on stinging nettle as a food source along the way. A review of Dr. James Duke's Phytochemical and Ethnobotanical database shows nettles to contain high amounts of calcium, magnesium, potassium, and phosphorus as well as trace amounts of nearly all the elements humans need to ingest to survive. (See appendixes C and E.) Joan and I eat nettles in soups, quiches, and casseroles—and it makes a delicious tea. I drink nettle tea nearly every morning all year round, harvesting and dehydrating the young plants in the spring to make tea late in the season and during the winter months.

There are several stands of nettles in my home landscape, and I use them for different purposes. The hoop house has a patch that comes to life in the middle of February, providing some of the earliest greens, which are much appreciated for fresh eating and used for tea. The old chicken coop has a patch that I let grow to full maturity and cut late in the year for use as a green manure when harvesting potatoes, as discussed previously. A nearby farm has a large stand that I cut and dehydrate for tea or use as a layer on the compost pile. Making fermented plant juice of nettles provides a shelf-stable, broad-spectrum mineral amendment that cannot be beat. This amendment is a staple in my gardening amendment toolbox. Growing around and in my garden, nettle plants send out lots of runners, making the species easy to propagate. I seldom dig out nettles from the garden but allow them to move around as they wish, growing among the other plants as a companion, to be used as a cover crop, or harvested for the dinner table.

Dandelions

Dandelions (*Taraxacum officinale*) are also very nutritious and tasty, and their long taproots pull minerals from deep below the surface and extend the tilth depth of the soil. I use the plant tops to make another great broad-spectrum mineral amendment. All parts of the dandelion are edible. The new young leaves in spring are a top-shelf early bitter green, the early flowers make delicious dumplings, and Joan and I dig the roots for a tea in the fall. Young flower buds can be harvested when still tight, dipped in batter, and sautéed to taste. The mature flowers can be pulled apart and added to a salad. Dig up some dandelion roots to chop, dry, and roast for brewing as a nutritive tea. I let dandelions go to seed in the garden, because I want them to be part of my garden every year. In addition to their uses for food and amendments, dandelions serve as part of the mixed cover crop that spontaneously grows in my garden. I use them as mulch, too, by harvesting the tops (leaving the roots in the ground in this case) and laying them on the soil surface.

Purslane

Purslane (*Portulaca oleracea*) is another extremely valuable weed. It is an abundant source of plant omega-3 fatty acid, the harvested greens add a wonderful texture to a salad, and it grows prolifically. This weed has very shallow roots, and thus it feeds a different stratum of the soil than dandelions do, for instance. If you do decide to pull some purslane out of the garden, don't leave it lying on the soil surface. Even small pieces of the plant can easily root themselves. I have chatted with many a gardener who has fought to remove this plant from their garden, unaware of how tasty and nutritious it is.

Taking note of plant root depth is important in terms of the new garden model, which recognizes that plants provide significant percentages of the nutrients formed during photosynthesis to the soil through the root system. Ensuring that there are roots present at various depths within the soil feeding different layers of soil biology becomes important. We can understand the benefit of allowing the deep taproots of dandelions, parsnips, or horseradish to grow adjacent to the shallow roots of nettles, purslane, or clover. Yet another insight into the wisdom of companion planting.

These are just some examples of how to take advantage of weeds that grow in the garden to improve the soil, and your own health. Some of my garden space has a thick covering of "weeds" in the spring, and most are there by my deliberate selection. As mentioned above, I occasionally cut back these wild cover crop areas with a string trimmer, aiming to set back the tall plants of early spring and allow the seeds underneath a chance to grow. I selectively avoid cutting down the plants that I want to go to seed. In this way, I maintain an extremely diverse cover crop that I do not have to plant. The crop comes up on its own each year, provides a diverse diet for soil biology at various depths, and becomes mulch at the end of the season.

Crushing Tool

A simple tool that is easy to make works well to pulverize seafood shells, animal bones, and some rock formations like schists into a fine powder that may be added to the garden as part of a long-term amendment strategy. Find two pieces of steel pipe, one of a larger diameter than the other, such that the smaller pipe fits into the larger pipe with some room to spare so it can be moved up and down. Be sure the smaller-diameter pipe is longer than the larger.

Weld a flat piece of steel onto the bottom of each pipe and grind the outside diameter of the smaller pipe's flat so that it fits easily into the larger pipe. The flat piece on the bottom of the larger pipe may be as large as desired; it will serve as the stand for the device. The two flat surfaces serve as robust parallel plates for crushing materials. When you've collected a good supply of shells or bones, set up the crusher and load the shells or bones into the larger pipe. Insert the smaller pipe into the larger and raise and lower it repeatedly to crush the material within. You may want to dry some material as described in the "Vinegar Extractions" recipe on page 129. The resulting powder may be stored for distribution as wanted at a future date. What a great way to complete the natural, sustainable cycles using local materials that are freely available and may otherwise end up in a landfill.

Two lengths of scrap pipe serve well as a crusher for pulverizing shells, bones, and some types of rock material.

CHAPTER 4

Data and Measurements

D ata can take the form of both numbers and observations, and I encourage gardeners to collect both kinds of data. The garden journal can be the home for day-to-day observations of plant growth and notes on planting techniques, amendments made, and much more. The journal becomes an invaluable reference, providing a record of the work done, a place to analyze the effectiveness of strategies employed, and a guide for what to try next season.

For a quantitative measure of crop quality, it's helpful to learn how to use a refractometer to measure the sucrose in plant sap. And for figuring out the nitty-gritty of how much of an amendment to apply, working the numbers of some simple mathematical formulas provides good answers. Because the amendment recipes presented here are often made from natural plant materials and contain minerals in plant-available forms, they can be applied with little worry about toxicity. However, with macrominerals, large amounts of some minerals may be needed, and in these cases getting the quantities right becomes important.

Two key categories of information guide decision making and calculations. The first is soil test results, which quantify the starting place—the quantities of minerals currently present in the soil. The second category of data is information about the quantities of specific minerals found in the amendments. In the case of rock dusts, this information may be obtained from the originating quarry. Purchased products should have this information on the label. For the amendments made from the recipes in this book, some of this data is available in appendix E and also can be figured out by extrapolation from James Duke's database of plants (as explained in "Weeds and Crop Plants" on page 113). Once the data is gathered, it's not so hard to calculate just how much of an amendment is needed to supply the correct amount of minerals, and I walk through several examples of these calculations in this chapter. The key is to recognize that multiple applications may be required and not to add too much at any one time.

Using a Refractometer

Plant health may be measured with a simple tool, the refractometer, which measures the sucrose (sugar) percentage in the plant sap or fruit. As the health of the plant improves, the percent of sugar in the sap increases, making the plant more resilient to pathogens and insect pressures. Macro- and micromineral availability of the right proportions, efficient photosynthesis, a highly functioning soil ecology, water, and sunlight all contribute to plant health. With improved photosynthesis efficiency and the production of complete proteins, excess energy is converted to produce lipids. This is seen as a thick, shiny layer on the leaf surface that must be penetrated by airborne pathogens in order to affect the plant. Add to this a highly functioning soil biology with minerals in the correct proportions, and insects, without enzymes in the stomach needed to break down complex sugars, are no longer a threat to a plant with a high complex sugar content—that plant is no longer a food source for these insects. Healthy plants with very high sugar content have the ability to thrive without the use of chemicals and are the food-medicine needed for human health. High-quality, healthy plants contain higher sucrose levels in their sap. Low-quality, disease-prone plants have low sucrose levels, and this may be measured with a refractometer.

The refractive index measurement device was established in the late 1800s for measuring and selecting the best grapes for making wine, because grapes with high sugar content are required to make good-quality wine. The refractometer is now used in many industries. There are no moving parts, and no batteries are required. It is a simple, inexpensive tool that is easy to use and provides relevant information about plant health and plant health trends.

The process of measuring the refractive index of plant sap or fruit juice is easy. Squeeze a drop or two of the liquid onto the glass surface of the refractometer and close the outer cover. Next, tap the cover

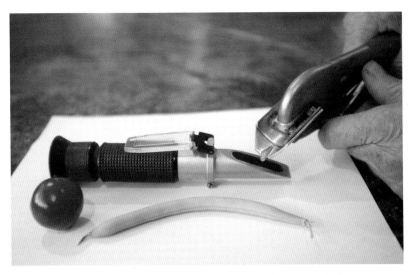

A modified vise grip squeezer makes it easier to apply extracted plant sap to the surface of a refractometer. I got this squeezer from Pike Agri-Lab Supplies.

gently until the entire surface of the refractometer is covered by sap with no air pockets present.

Look into the device. The refractive index is interpreted as the place on the scale where a color change can be seen. Often the color transition is from white to blue. The scale is calibrated such that each increment represents a 1 percent increase of sucrose in the sap. Each increment, or percent sucrose, is referred to as one Brix. If the color transition is observed at 5.8, for example, then the refractive index of the sap is 5.8 Brix, which equals 5.8 percent sucrose. The transition between the two colors is not always definitive; it may be blurred. When you record refractometer values in your garden notebook, make note of both the nominal value and the width of the blur.

You can use a refractometer to evaluate the nutrient content of fruits and vegetables purchased from a grocery store or farmers market. For example, blueberries have a great reputation as a health food because they are reputed to be high in antioxidants. But not all blueberries have the same level of antioxidants. One way to measure the nutritional quality or health of a blueberry is to measure the refractive index of the fruit sap as described above. Appendix F provides a listing of refractive index ranges for selected plants as compiled by Dr. Carey

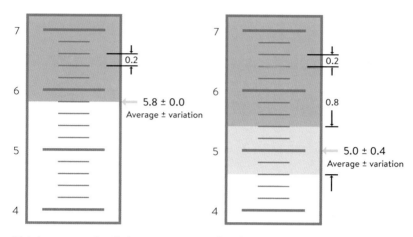

This is an example of what you may see when looking into a refractometer that contains a sample of plant sap. The values to take note of are the nominal value and the width of the blurry transition between colors (5.8 ± 0 and 5.0 ± 0.4).

Reams in the 1940s. On this scale a poor-quality blueberry is listed as 10, while an excellent one is listed as 20. What is the quality of the blueberries you had for breakfast?

Record values measured in your garden notebook for long-term evaluations. Calibrate your tongue by measuring the refractive index of a fruit and tasting it at the same time. The measurement of sap refractive index has high variability, but is an excellent relative comparison tool.

If you want to measure the quality of a plant that is not included on Dr. Reams's list, start your own database. Let's say you're a radish fan. Measure the refractive index of the flesh of all the radishes available to you. As you take the measurement, taste the radish, too, and evaluate the flavor. The goal is to establish measurement values of poor, good, and excellent radishes and to calibrate your taste buds at the same time. With 25 to 30 measurement values, the standard deviation and range of radish health variation may be established and used for evaluating the quality of radishes going forward.

This measurement tool can be used for many purposes. Use your imagination!

- Compare the produce purchased from different farms or stores.
- Compare organic versus non-organic fruits and vegetables.

- Compare this year's crop with last year's, and year by year you can gauge improvements.
- Evaluate the efficacy of amendments applied to plants by applying a foliar spray early in the morning to some of your blueberry bushes but not others. The value of the refractive index should go up within hours of a foliar spray application if the plants like it. Measure the refractive index of each group a few hours later and the next day to evaluate the response. Record this information in your garden notebook.
- Measure the refractive index of particular plants at two-hour increments during the course of a day. If the refractive index does not change during the day, it is an indication of a boron deficiency.

In *Exploring the Spectrum*, Dr. Philip S. Callahan notes that lower Brix levels suggest low phosphate levels while appropriate Brix levels, according to plant species, indicate relative immunity to bacterial, fungal, and insect attack.

Nature understands this plant sap sucrose content scale as well. I have always been intrigued when I visit a plant nursery where many rows of hanging plants loaded with flowers in baskets are on display, but no pollinators are buzzing around any of the blossoms. How could this be? A honeybee will not visit a flower that has a percent sucrose level less than 7 Brix. The nutrient content of low-Brix flowers seems not worth the effort to nurture the honeybee.

Plants become resistant to insect damage when Brix levels are above 12. Insects have built in refractometers. Perhaps they "see" unhealthy plants using parts of the electromagnetic spectrum out of the visible range. In fact, they do! See *Tuning in to Nature* by Philip S. Callahan.

Refractometers are easily available for order online from a range of suppliers. The desirable range of measurement for most plants is 0 to 35 Brix. Garlic, which can have Brix values near 50, is an example of the ranges possible in nature. No wonder garlic tastes so good!

Conducting a Soil Test

Soil testing is a simple, straightforward process. An important first step is to identify a testing lab that will conduct a test that includes all the microminerals and trace minerals of interest. I have been using Logan Labs in Lakeview, Ohio; it provides an analysis that includes mineral analysis, exchange capacity, cation proportions, and more (www .loganlabs.com). Soil tests need not be conducted every year, making this a reasonable investment, especially if no information about the mineral content of the soil is available.

In order to evaluate soil mineral content, the lab staff rinses samples in a weak acid solution. The acid solution simulates the weak acids that the soil biology produces in order to break down minerals (rock particles). This test is a way to estimate the availability of minerals in the soil. Acids of different strengths are used to rinse the soil, and the test names are a way of differentiating the strength of acid used. The test known as the Mehlich 3 test provides a reasonable measure of plant available minerals in the soil. More recently a testing method called a Haney soil test has become popular. Discussing these test methods with a specific lab will shed light on which may be best for a particular situation. A lab should offer detailed directions on how to conduct a soil test and the nature of its analysis. Look for this information on the testing lab's website. Once you've chosen a lab, it's a good idea to keep using it over time to limit any sources of test variation when comparing results year after year.

Another way to reduce variation is to collect soil samples for testing at the same time every year. It is interesting to note that soil mineral proportions change over the course of a year, and that the fall represents a low point in soil mineral content. This is a lovely concept to ponder, because it brings to light the power of soil biology over those cold winter months and early spring working to digest available minerals for next year's plants to ingest. Yes, the soil digestive system continues working even in the depths of winter. Conducting a soil test in the fall captures this low point of the mineral availability cycle, and it's a time of year when other garden chores may have subsided.

In general, collect samples from the top 6 inches (15 cm) of the soil, which is (theoretically) equivalent to the depth of the root zone of many crops. This 6-inch depth is also relevant when doing calculations of amendment needs, as I explain in "Garden Math" on page 104. There is a special device one can buy that collects perfect "cores" of soil. Another easy way to collect a sample is to use a trowel. Before digging, be sure the tools used for gathering, mixing, and packaging are clean and not contaminated by rust or other contaminants. Scrape away any plant or other debris from the surface of the soil and dig the hole a little over 6 inches deep. Remove a sample about 2 inches (5 cm) thick from the side of the hole, slicing down about 6 inches. Be sure that the sample does not include any material other than soil.

Gather several samples in this manner from various spots in the garden. As a rule of thumb, collect five or six samples within a 50-by-50-foot (15 × 15 m) area. The goal is to determine the average characteristics of the garden's soil overall. Next, thoroughly mix all the individual samples together in a clean bucket. Once mixing is completed, remove 1 to 2 cups of soil from this bucket. This represents an average of the top 6 inches of soil in the garden. Package this up and send it to the lab for analysis. And remember, even more detailed instructions for collecting samples should be available through the testing lab.

The first time the soil is tested, the results may be disheartening, but fear not. Even in a new garden that has some soil imbalances, the harvest of freshly picked produce will still be more nutritionally rich and flavorful than store-bought produce that was picked long ago and transported hundreds or thousands of miles. Also keep in mind that it may take longer than one year to correct many soil mineral deficiencies. It may take several years. The same may be true for excesses—it may take several years before those levels decline to an optimal range. The important thing is that objective information is now available to develop both long- and short-term amendment strategies. Just imagine how good those tomatoes will taste next year!

In addition to soil testing, I recommend trying to figure out what the mineral content of the soil might be by observing what grows there, especially the common weeds, as suggested in "Weeds" on page 91.

Amendment Analysis Results

The more I worked with the amendment recipes in the book, the more I wanted to know what minerals were present in them, and thus I conducted a mineral analysis on many of them. This analysis assessed the concentration of many minerals that are immediately available and accessible to plants and the soil solution by using deionized water to extract the sample to be analyzed from each amendment. I found that all the amendments contain a broad spectrum of the constituents evaluated, but that the proportions vary. The amendments contain many kinds of ions as well, as evidenced by the increase in electrical conductivity when amendments are added to a bucket of rainwater. The electrical conductivity of a 4-gallon (15 L) bucket of rainwater increases as much as 0.1 mS/cm when 2 tablespoons of some kinds of amendments are added to the bucket at a dilution ratio of 500:1. The amendments also contain enzymes and other higher-order compounds. All these amendments are plant-based, containing minerals and compounds in plant proportions. All these constituents are water-based and in forms that plants can use directly when they are applied as either a foliar spray or a drench.

The analysis of some of the amendments made from these recipes is summarized in appendix E. This analysis may be reviewed in conjunction with Dr. James Duke's Phytochemical and Ethnobotanical online database. Dr. Duke's data provides ranges of constituents within each plant, offering an indication of the standard deviation between measurements. It is always interesting to compare results. Establishing a statistically based data set of constituents within amendments made per these recipes would be a great project.

I share this data I've collected to show how diverse and powerful these amendments are, and to help you understand the wealth of materials available at your fingertips to help feed your soil and plants. Each extraction method—vinegar, water, sugar fermentation, and leaf mold fermentation—produces amendments with unique characteristics.

Appendix E includes data for the content of a dandelion amendment fermented with organic brown sugar and also for a dandelion amendment fermented with leaf mold. The leaf mold fermentation has a significantly lower concentration of almost all constituents, but more time fermenting in the bucket may change this result. However, it has the advantage of being less expensive to make, because organic brown sugar must be purchased and leaf mold is free for the taking. And there's the indirect environmental cost of producing and shipping that organic brown sugar from its point of manufacture to the local store or food co-op. There is something to be said for filling a bucket with dandelions and rainwater, adding a handful of leaf mold from the local woods, and letting it brew for a time. It's a mighty easy and convenient way to make a broad-spectrum amendment product. On the other hand, the sugar-fermented product will be complete in about a week and be shelf-stable for use until it is used up. Another way to think about these two amendments is to consider what enzymes and other complex compounds each may contain. This is much harder to measure, and this is why experimentation with them becomes important.

I'm intrigued by the fact that organic apple cider vinegar can further extract minerals from the residues of other processes. For example, in one experiment I used organic apple cider vinegar to extract the residue left after making fermented plant juice of dandelion. I extracted the residue four successive times, and all four extracts contained more than twice the concentration of manganese and over an order of magnitude more zinc than did the original fermented plant juice of dandelion. I also extracted the residue from making a batch of fermented fish amendment. The extract showed concentration increases of phosphorus, calcium, boron, manganese, and silicon compared with the original amendment. As the concentrations of some minerals diminish, the concentrations of others may increase. Further study is required to ensure that these methods are repeatable, but this seems to be a way to establish high concentrations of specific minerals to augment a broad-spectrum amendment. Note also that a little goes a long way. A quart of this concentrated material may be diluted 1:1000, giving 1,000 quarts (L) of a specific concentration of zinc, for instance.

A most fascinating observation from this data is that every amendment contains at least some tiny amount of nearly all the minerals plants require. This further supports the ideas that the best amendment for a plant is the plant itself. The mineral proportions of an amendment made from blueberries are exactly the mineral proportions that a blueberry needs. Using the data in appendix E, it can be seen that some concentrations are higher than others. Scan the data for products high in boron, cobalt, molybdenum, or sulfur. Reviewing this data reminds me why a balanced diet is so important. I think I will eat a peach right now!

I have not spent much energy analyzing the biological amendments I have made. Biological analyses are expensive and take time. Also, because of how much we do not yet understand about the varieties of bacteria, fungi, and archaea in the soil, it is not clear whether any analysis available today could quantify all these organisms in a meaningful way. It is always important to have something else to do. Perhaps someone who reads this book will be inspired to undertake this line of investigation.

Garden Math

Here are tools that may be used to determine how much of a mineral to add over a certain area when addressing a mineral deficiency identified in a soil test report. The trick is to be able to convert between pounds per acre and parts per million. For those who just want the answer and have no interest in the reasons why, the basic rule is that 2 pounds of a mineral per acre is equivalent to 1 part per million (ppm) when distributed evenly over 1 acre of land 6 inches deep. For those who like to know the why, here is the math that explains the rule.

A starting point is needed. The volume of soil in a typical root zone depth of 6 inches and 1 acre in area weighs 2 million pounds. One part per million is simply 1 divided by 1 million. So for 1 acre of soil 6 inches deep weighing 2,000,000 pounds, 1 part per million of the soil by weight is 2 pounds.

$$\frac{2{,}000{,}000 \text{ pounds}}{1 \text{ acre}} \times \frac{1}{1{,}000{,}000} = 1 \text{ part per million}$$

$$\frac{2 \text{ pounds}}{1 \text{ acre}} = 1 \text{ part per million}$$

In practical terms, minerals are usually applied only on the soil surface and not evenly distributed in the entire volume of soil, but the thinking is that after some time the minerals will become distributed throughout the 6 inches of the root zone. The reality is that mineral distribution depends on many factors, including weather, mineral solubility, the dynamics of the soil solution, and the functioning of the soil ecosystem, to name a few.

The next step is to understand how to apply this relationship with different parameters using a mathematical equation. The basic mathematical idea is that whatever is done to one side of an equation must be done to the other. We start with the idea that 2 pounds per acre is equivalent to 1 part per million.

Example 1

Double the number of pounds added to the acre of land. How many parts per million is this?

To double the number of pounds added to the acre of land, multiply both sides of the equation by 2. So 4 pounds of material added to the acre of land is equal to 2 parts per million of that material on the acre of land.

$$\frac{2 \text{ pounds} \times 2}{1 \text{ acre}} = 1 \text{ part per million} \times 2$$

$$\frac{2 \times 2}{1} = 4 \text{ (pounds of material) equals } 1 \times 2 = 2 \text{ ppm}$$

Thus, 4 pounds of material equals 2 parts per million.

Example 2

How many pounds of a mineral is required in order to add 3 parts per million of that material to a garden 50 feet by 40 feet?

$$\frac{2 \text{ pounds} \times 3}{1 \text{ acre}} = 1 \text{ part per million} \times 3$$

Six pounds of the mineral is needed on an acre in order to have 3 parts per million. Next, convert our garden area from square feet to acres (1 acre = 43,560 square feet).

$$40 \text{ ft} \times 50 \text{ ft} = 2000 \text{ ft}^2$$

$$2000 \text{ ft}^2 \times \frac{1 \text{ acre}}{43,560 \text{ ft}^2} = 0.046 \text{ acre}$$

Equate the two area ratios and solve for X pounds to be added to the garden. Remember from above that 6 pounds of the mineral is required on an acre in order to have 3 ppm.

$$\frac{X \text{ pounds}}{0.046 \text{ acre}} = \frac{6 \text{ pounds}}{1 \text{ acre}}$$

$$X \text{ pounds} = \frac{6 \text{ pounds}}{1 \text{ acre}} \times 0.046 \text{ acre}$$

$$X = 0.276 \text{ pound} = 4.4 \text{ ounces}$$

So 4.4 ounces of the mineral added to a garden 40 feet by 50 feet is equivalent to adding 3 parts per million of this mineral to the garden.

As a final step, consider the addition of a compound with multiple elements and how to evaluate how much of each element within the compound is being added.

Example 3

If 35 pounds of gypsum were added to 1 acre of land, how much calcium and sulfur are actually added in parts per million?

The amount of an element added is the weight of the compound, in this case gypsum, times the ratio of the atomic mass of the element within the compound divided by atomic mass of the entire compound. The chemical composition of gypsum is $CaSO_4 + 2H_2O$.

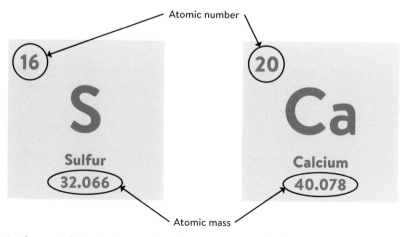

In the periodic table the number above an element's chemical symbol is its atomic number. The number below the element name represents its atomic mass.

Table 4.1. Calculation of Atomic Mass of Gypsum ($CaSO_4$-$2H_2O$)

Element	Quantity of Element	Atomic Mass	Atomic Mass of the Element in Gypsum	% (by weight) of Element in Gypsum
Ca	1	40	40	40/172 = 23
S	1	32	32	32/172 = 19
O	6	16	96	96/172 = 56
H	4	1	4	4/172 = 2

The atomic mass of this compound is 172. (See periodic table extraction on page 106 and the table calculating the atomic mass of gypsum on this page.) The atomic mass of calcium is 40, and there is a single atom of calcium in the compound. The atomic mass of sulfur is 32, and there is a single atom of sulfur in the compound. The atomic mass of oxygen is 16, and there are 6 atoms of oxygen in the compound. The atomic mass of hydrogen is 1, and there are 4 atoms of hydrogen in the compound. The atomic mass of the entire compound is 172.

(A) How much calcium is added to an acre of land if 35 pounds of gypsum added?

$$35 \text{ pounds of gypsum} \times \frac{40 \text{ (atomic mass calcium)}}{172 \text{ (atomic mass gypsum)}} = 8.1 \text{ pounds calcium}$$

Thus, 8 pounds of calcium (round 8.1 to 8.0) are added if 35 pounds of gypsum are added to 1 acre of land. Using the math in example 1, for an acre of land 6 inches deep, 8 pounds of calcium equals 4 parts per million calcium. So adding 35 pounds of gypsum to an acre of land is equivalent to adding 4 parts per million calcium to that acre of land.

Using the math in example 2, 35 pounds of gypsum added to ½ acre of land doubles the concentration to 8 parts per million calcium on that land.

(B) How much sulfur is added to an acre of land if 35 pounds of gypsum added?

$$35 \text{ pounds gypsum} \times \frac{32 \text{ (atomic mass sulfur)}}{172 \text{ (atomic mass gypsum)}} = 6.5 \text{ pounds sulfur}$$

Thus, 6.5 pounds of sulfur are added if 35 pounds of gypsum are added to 1 acre of land. Using the math in example 1, for an acre of land 6 inches deep, 6.5 pounds of sulfur equals 3.25 parts per million sulfur. So adding 35 pounds of gypsum to an acre of land is equivalent to adding 3.25 parts per million sulfur.

Using the math in example 2, 35 pounds of gypsum added to ½ acre of land doubles the concentration to 7 parts per million sulfur on that land.

Applying the calculated amount as several partial applications over a period of time should be considered. It is easy to add more, difficult to add less once it is in the ground.

A Garden Notebook

Start a garden notebook and record all that comes to mind. Record where you collect ingredients such as plant materials or shells. When you make an amendment, list all the ingredients used, when the amendment was made, and how long you left the amendment sitting before decanting it. Record when you apply an amendment, the dilution rate, and how much of the diluted solution was applied to what number of plants or number of square feet. Make notes about what worked and what didn't, conditions of application such as weather, and time of day.

Use this notebook to record other aspects of your gardening, too. Record what you plant and when, biodynamic calendar day type (if you use the biodynamic calendar), harvest dates, crop size, flavor quality, and Brix reading. List the weeds that grow in your garden throughout the season, including the time they sprout and when they flower. Record soil temperatures and air temperatures in different locations around your landscape. Sketch garden layouts and rotations. Write about how you feel when you're in the garden, or about anything that comes to mind. Reflect on the natural processes around. The act of writing things down helps to remember and reflect year after year. This is not a single event, but an annual activity that will be repeated for a lifetime. Print and clip articles of interest. Tuck them into your notebook and watch it grow. It may quickly turn into several notebooks on different topics. Simply documenting what you see when you spend time outdoors is a wonderful meditation to be reflected on, nurtured year after year.

The making of the four stages of IMO is an excellent example of the importance of documenting the process. The duration that rice is left in the trap box, and the associated soil temperature and rainfall during that time, will affect the degree and thickness of white fuzz that forms on the top of IMO #1. If IMO #2 is left to ferment too long, the result will be a useless gruel. I often conduct a soil test of the soil I use when making IMO #4 as well as a soil test of the final product to confirm the addition of specific minerals added and their amounts. Document how much water is added at each step for IMO #3 and IMO #4. The amount of water, the amount of sun, and the covering of the pile all affect the liveliness and longevity of the biology within it. By documenting what is added, how much, and when, a guide for the next time around is provided.

It is also important to discuss results with others, to share and record the collective experiences of the outdoor world. One of the most common things heard in any garden discussion is, "What a strange year it has been!" In fact, every season is different, and the differences between seasons are becoming more varied. Some plants do well and some not so well in any given year, as do the populations of insects, rodents, and all the animals in the food chain. These are important discussions. I am reminded of the year that temperatures in February dropped to –15°F (–26°C) a couple of times and most of the peach blossoms in the state of Connecticut were killed there was no peach crop that summer. Suppose that you had invoke a new protocol for amending your peach trees that year. Without conversations with others who were growing peaches leading to the realization that all peaches were affected by the cold weather, you might have concluded that your new protocol was no good, but by documenting and discussing with others you can reach a more accurate conclusion.

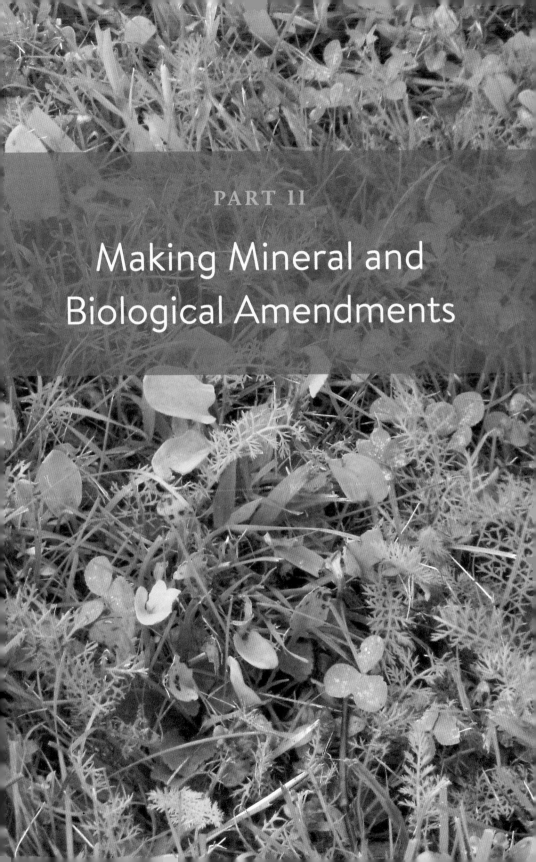

PART II

Making Mineral and Biological Amendments

CHAPTER 5

Raw Materials

S ome of the raw materials I use to make garden amendments are very basic—plant parts, good water, vinegar, raw milk, leaf mold from the woods. Many mineral amendments are available in the backyard and at local natural areas and businesses; they need merely to be recognized. Among these are rocks, muck, plants, seafood shells from restaurant kitchens, bones from butcher shops, and fish wastes from fishing docks. Some imagination is required to recognize these resources, but many can be incorporated into a long- or short-term soil mineralization program with applications in small amounts throughout the year. Their power after sitting in a bucket to ferment is nothing short of amazing. The raw materials listings that follow reflect many of the type of resources available all over the world or that may be used to close waste gaps.

Weeds and Crop Plants

By selecting particular types of plants growing around gardens and farms and fermenting the leaves and stems, it's possible to create amendments that contain the trace minerals needed to address plant and soil micromineral deficiencies. As explained in part 1, this is because plants have the ability to accumulate minerals in proportions that often differ from those found in the soil solution. We can use this knowledge about the types of minerals that a specific plant accumulates when we make amendments. An amazing resource for researching which minerals individual plant species accumulate is Dr. James Duke's Phytochemical and Ethnobotanical database. Dr. Duke was a renowned botanist who developed the database during his career with the USDA Agricultural Research Service. This database provides nearly 50,000 listings, including listings for more than 2,300 plants. Each plant listing details the mineral constituents, and a range of values is given for each constituent. Appendix C is an example of the type of detailed data available in this database; the appendix shows the mineral constituents of dandelions, nettle, and sassafras. The database itself can be freely accessed online at https://phytochem.nal.usda.gov/phytochem/search.

Compare these values with the results of a laboratory soil sample analysis to identify which plants could be useful for making amendments that will help supplement the existing soil solution of your growing area. The desired levels of microminerals like boron, copper, zinc, cobalt, molybdenum, and selenium are on the order of single parts per million, and thus the mineral amendments described in this part of the book are a viable vehicle for supplementing soil deficiencies.

Here's an example of how to make use of the information in the database. Based on the information in appendix B, "Optimal Soil Mineral Amounts," we know that the desired level of selenium in the soil is roughly ½ part per million (0.5 ppm). According to Dr. James Duke's database, stinging nettle leaf contains 0.4 to 2.2 ppm selenium, and thus it should be a viable local, sustainable, regenerative source of selenium in a plant-available form. To confirm this, I had a batch of fermented stinging nettle juice I made analyzed by a testing lab. The analysis showed that the fermented juice contained a selenium concentration of 1.03 ppm. The fermented nettle juice also contained a broad spectrum of other minerals and complex compounds, all in forms that plants can absorb directly through their leaves and stems when the diluted juice is applied as a foliar spray or via the soil–root ecosystem when applied as a drench. Appendix E provides examples of the elemental analysis of many of the amendment recipe products described in this book, and it will serve as another useful reference as you make choices of which plants to use when making amendments.

Using a single plant type is recommended when making an amendment because that allows for more accurate prediction of the unique composition of mineral constituents the amendment will contain. (The biological constituents of each amendment will be unique, too.) This also allows for amendment repeatability when identifying one that works well in a specific situation.

A Homeopathic Approach

Homeopathy is a system of medicine that treats conditions with minute amounts of natural substances. The products of the recipes in this book have a homeopathic aspect because they provide plants with trace minerals and other organic compounds that may be present only in very small concentrations—less than a single part per million. When using these products, the first step is often to dilute the extracts by as much as 1:1000. This is similar to the dilution levels of the remedies used by homeopathic medicine practitioners. The idea that extremely small amounts of something may produce profound effects is an underlying principle that is extremely important to consider when working with such complex systems as life.

It's not always necessary to consult the database to figure out which plants to use to create a suitable amendment. As explained in chapter 2, the best source material for making an amendment to support a particular species of plant may be plant parts of that same species, with slight adjustments. For example, the fruit of a plant includes the quintessential nutrients that plant requires in order to reproduce. Rather than trying to determine what a plant needs, why not make use of the plant's own wisdom and resources! Make a fermented plant juice using the rotten, unripe, insect-infested, or frost-killed fruit from the plant in the fall. Store the ferment over the winter and use it to feed that crop the following year. Ferment carrot greens to feed your carrot crop, beet greens to feed beet plants, insect-ravaged or rotten peaches to feed peach trees—you get the idea. Adding a vinegar extraction of eggshells to the fermented plant juice would provide beneficial calcium as well. Adding a vinegar extraction of cow bones will add calcium and phosphorus. Note that both these vinegar extractions have their own spectrum of minerals as well.

Local Rocks and Soils

Minerals abound in rocks and soils all around. Calcium and magnesium can be found at a lime quarry, silicon and maybe manganese from a basalt quarry, paramagnetic materials from basalt or granite quarries, calcium and sulfur from gypsum, boron from borax, and silicon from diatomaceous earth. Other sources of local minerals are the silts that build up on riverbanks after spring flooding and the muck around swamps and bogs.

Rock Dusts

Some types of rock formations, such as basalts, contain trace minerals, and rock dusts can be used for long-term mineralization of soil. Basalt and granite quarried rock dusts may also have paramagnetic properties. Paramagnetic properties of soil are becoming more widely recognized as an important characteristic, in part because microorganism health and the flow of energy between plant and soil are affected by the magnetic susceptibility of the soil, as explained in "Improving Energy Flow in Soil" on page 17.

Consult a local geological survey map to identify rock formations in the local area. These are the geological survey maps that identify the minerals below the ground surface, not the ones used by hikers that show elevation, rivers, and roads. These maps will identify the rock formation types in the area. Comparing this information with the analysis results from a soil test will identify whether the rocks in the area have minerals needed for the garden. With the knowledge of local rock types and soil needs, next identify rock quarries along the rock formations of interest. The very fine dust that accumulates as a result of crushing rocks is a quarry waste product. Quarries must dispose of the dust on an ongoing basis, and it is often available free for the taking. Quarries frequently conduct analysis of the rocks crushed in order to advertise the mineral composition of their product. This information is usually provided upon request. If analysis is not available, conduct an analysis of available materials before applying them to your soil or using them in a recipe. It's important to determine that a particular rock dust contains appropriate minerals for your soil and plants before application.

Another source of soil mineral composition information is the *Element Concentrations in Soils and Other Surficial Materials of the Conterminous United States*. This reference can be used to identify the general soil mineral distributions of many minerals anywhere in the United States. The maps provide information about the mineral content of the soil at a state level. For more detailed local information about the soil mineral composition in the backyard, seek out bedrock geological survey maps for your vicinity available from the United States Geological Survey (USGS). These sources can be used to evaluate general local rock formations and the minerals that may be common in the soil. Using this information in conjunction with the location of local quarries will help identify what mineral types are available in your area and a general understanding of mineral composition of soils in that area.

Visiting a quarry and asking about the mineral composition of the rocks mined and the availability of the dusts is interesting and informative. Once mineral composition is understood, comparison with a soil test allows informed decisions about need and application.

It takes time for rock dust particles to become assimilated into the soil, because they must be digested by the soil biology in order to enter into the soil solution. Rock dusts can also be applied to compost piles or to a pile of IMO #3 when making IMO #4 to facilitate their digestion.

Apply rock dusts to the soil in late fall through early spring in multiple small applications. Sprinkle rock dusts over growing areas before and/or after applying mulch. Applying a biological amendment, an organic carbon source, and rock dusts all at one time is an excellent strategy because it provides biology and food and housing for that biology, which in turn will enhance digestion of the rock dust and its assimilation into the soil solution.

Silts and Clays

Silts and clays from local bogs, swamps, ponds, and streams that flood in the spring provide minerals and enhance the soil's exchange capacity. Be mindful of who owns these places and any laws prohibiting removal. Extracting small amounts makes this practice sustainable. Conducting an analysis of this material provides an understanding of what minerals are present, and assures that unwanted elements such as heavy metals are not. It is easy to add silt or clay to a garden area, but difficult to remove it! It's a good idea to collect samples of silt or clay and send them to a lab for analysis. Test results provide a basis for making informed decisions about whether to apply the material and how much.

Silts and clays are used for long-term mineralization and for building a soil's cation exchange capacity. Apply to the soil in late fall through early spring, or apply the material to a compost pile. Try adding small quantities of a mineral-rich silt or clay to planting holes, or spread it when applying mulch. As explained above for rock dusts, it's a good practice to use biological amendments in conjunction with silts and clays.

Ocean Products

If you are lucky enough to live near the ocean, there is a bounty of free raw materials at your disposal. Seaweeds, shells, and salt water itself are all good amendment sources with broad-spectrum mineral composition, biological sources, and other compounds. Oyster, clam, and crab shells are good sources of calcium and trace minerals. Salt water can be diluted with good water 1:50 and applied to the garden or when making IMO #3 or IMO #4. Seaweeds can be fermented, extracted, used as mulch, or added to a compost pile. Shells can be cooked and extracted with organic apple cider vinegar or crushed and applied to the soil, compost pile, IMO #3, or IMO #4. Shells crushed and mixed with rock dusts make interesting mineral amendments. These are all great reasons to go to the beach at any time of year. An easy local alternative to the ocean available to most are the shells discarded by a local restaurant or seafood distributor. These shells otherwise often end up in a landfill. Close the waste gap.

Raw Milk

Raw milk, rich in biology and nutrients, supports life from birth. It is the quintessential nutrition source. The biology in raw milk inoculates the gut, soil ecology, or plant surface. The nutrients nourish the animal, plant, or soil biology. Raw milk is an effective mineral and

If free or inexpensive raw milk is available, it can be diluted with good water for use as a great mineral and biological amendment.

biological amendment that can be applied to the ground, to plants, or to a compost pile as a foliar spray or drench diluted 1:10 or up to 1:30 with good water. Raw milk can also be used to make lactic acid bacteria. (See "Lactic Acid Bacteria" on page 154.)

Only 11 states allow the sale of raw milk (from cows) in retail stores. Twenty states prohibit the sale of raw milk altogether. What's up with that? The answer is quite complex, with people arguing passionately on both sides of the issue. If you're not familiar with the availability of raw milk in your area, ask around to find out whether there's a local source and the means to get it.

I use any raw milk that has gone by for gardening purposes to avoid wasting this valuable product. I add small amounts of spoiled milk to the compost pile, or put it into a foliar spray or add it to a soil drench. I have not collected any quantitative data to report on its efficacy.

CHAPTER 6

Amendment Recipes

Making sustainable gardening amendments is not like follow-ing a highly refined recipe crafted by a professional chef. This is not baking, but more like making a soup and fol-lowing the brief notes from your grandmother for one of her favorite recipes crafted through experimentation and instinct. I have written these recipes for the home garden scale, but all of them can be scaled up to any level desired.

These amendments can be made in any kitchen using simple house-hold tools and ingredients (jars, crocks, sieves, sugar, vinegar), and in small batches that allow for easy experimentation and customization. The quantities listed in the recipes are offered as a general guide, not a rigid protocol. Thankfully, life is tolerant and tenacious, allowing for significant variations.

Experimentation is a good thing with these recipes, so don't fret if the materials listed are not precisely at hand. Use what is available to you. The same goes for equipment. A few buckets of anywhere from 2 gallons (8 L) to 20 gallons (80 L) in size will come in very handy, as will stoneware crocks. Glass jars with tight-fitting lids, such as canning jars, work well for storing and making the amendments.

WATER EXTRACTIONS

Amendment type: Biological or mineral, depending on length
of extraction period; shelf-stable
Raw materials: Rainwater; plant leaves, flowers, stems
Content: Broad-spectrum minerals and other compounds;
biology (from plant surfaces)

This is the simplest and oldest of amendments and one of the cheapest to make. Plants, often common weeds, placed in a bucket with good water added develop into both mineral and biological amendments. As I've emphasized earlier in the book, each species of plant has a unique mineral content. Nettle and dandelion are especially useful choices for extraction because they are the multivitamins of the plant world.

In addition, living plants are covered with hundreds of thousands of indigenous microorganisms. During the extraction process, these microorganisms become suspended in the water in the bucket and digest the plant material over time. Once you gain an understanding of the weeds in your area, you can quickly and easily make effective broad-spectrum mineral and biological amendments by water extraction. The only downside of water extraction is that the smell quickly becomes extreme.

This recipe can be scaled up to any level and can be made at most times of year when plants are available. An example of the minerals measured in the liquid strained from a water extraction of comfrey is provided in appendix E. Remember that Dr. James Duke's Phytochemical and Ethnobotanical database is a great resource for learning about the unique mineral proportions of individual plant species. (See "Weeds and Crop Plants" on page 113 for more about this database.)

Materials

Bucket

Plant residue (leaves, fruit,
    stems, roots)

Rainwater or other good water

Clean piece of cloth

Board, large enough to cover
    bucket top

Instructions

Step 1. Gather material of a single plant type from your garden or another clean source. (Don't gather material along the side of a busy road, for instance.) Put the material in the bucket loosely. There is no precise right amount of plant material to use; filling the bucket about three-quarters full works well. Fill the bucket with rainwater or another type of good water.

Step 2. Cover the bucket with the cloth and then the board in order to keep insects and other debris out of the extraction.

Step 3. Store the bucket out of the sun in a well-ventilated area. If you plan to store it long-term, choose a spot that will remain above freezing during the winter months.

Uses

During the first two to three days, you can strain and dilute the liquid in the bucket one part to ten (1:10) or as much as one part to thirty (1:30) with additional good water and use it as a biological inoculant by applying it as a drench on the soil or compost pile or as a foliar spray on leaf surfaces.

After a few days, the biology present in the bucket begins the decomposition process, and the minerals in the plant materials are distributed into the water. The pH of the solution will fall to about 5.5 after about a week. If you decant the extract at this point in time, the low pH allows for stable storage options. Be sure to label the extraction.

Step 1

Over time this decanted solution will change from acidic to basic, with pH values of 7.0 to 10.0. This change in pH was an unexpected and fascinating part of this process that is worthy of more study and offers more possibilities for experimenting with different uses of this basic amendment.

You can strain out the plant residue by pouring it through a sieve and dilute it right away to apply as a mineral amendment. Or you can store the strained liquid in glass jars with tight-fitting lids for future use as a mineral amendment. Be sure to label the container with the date, the extraction process used, and the ingredients. The plant residue may be further extracted using organic apple cider vinegar, distributed on the compost pile, or placed around the base of a choice perennial plant in your landscape.

Note that the quantity of minerals in a water extraction is relatively dilute compared with other fermentation methods such as Fermented Plant Juice (see page 134).

If you leave the materials to extract longer than about 2 weeks, the contents will begin to smell terrible rather quickly. Don't be afraid of bad smells—the soil ecology does not mind the stink, but if the bucket is stored close to where people are living, some may protest! Decomposition will continue for months, moving more of the mineral content into the water menstruum over time, but the smell may become overwhelming. The use of leaf mold (see "Leaf Mold Fermentation" on page 144) will greatly reduce the odoriferousness.

RECIPE AT A GLANCE

- ▸ Put plants in bucket with good water, cover, and let sit.
- ▸ For first 2 to 3 days, strain and dilute 1:10 or up to 1:30 and apply to soil as biological inoculant.
- ▸ After approximately 1 week, pH will drop to about 5.5. At this stage, liquid may be strained into glass jars, covered with tight lid, and stored for future use as mineral amendment if desired.
- ▸ Alternative: Allow decomposition to continue indefinitely in bucket if you can stand smell. As needed, strain and dilute 1:10 or up to 1:30 and apply to plants as mineral amendment.

APPLE CIDER VINEGAR

Amendment type: A shelf-stable menstruum used for making
mineral amendments
Raw materials: Apples, water
Content: Weak acid

Almost anybody can make bad wine. Vinegar is simply a batch of wine
gone bad. Homesteaders find many applications for apple cider vinegar, and it is very easy to make your own. For garden amendments,
apple cider vinegar can be used to extract minerals from shells,
bones, or other local materials. Its acidity breaks down mineral
constituents. Making apple cider vinegar with apples that would
otherwise be thrown away closes a waste cycle in a satisfying way.
Drops from a tree in your backyard or in a local field will work fine.
Many old apple trees that have been ignored and left untended still
put out quantities of fruit. If you collect apples from a farm, keep in
mind that there may be pesticide, herbicide, and fungicide residues
on the fruit. Many of these chemicals are not water-soluble and will
not simply wash off in water. Do you want to introduce them into
your growing environment?

It takes 2 to 3 months for apples to ferment into vinegar, depending
on ambient temperatures, and they should be kept out of the sun in
a well ventilated space and undisturbed during that time. If you are
planning to make a large batch, you may want to position the container
in the right environment for brewing before you add the apples and
water. When the pH drops to 5 or lower, the liquid is vinegar and may
be decanted. Once I set up a batch of apples to ferment, I generally
forget all about them until one day I walk by the crock and wonder
how things are doing. I discover that the months have passed by and
the contents are ready for decanting.

Materials

Unsprayed apples
Glass jar or crock
Good water
Plate or flat rock (optional)
Clean piece of cloth

Board, large enough to cover
bucket top
pH paper
Strainer

Instructions

Step 1. Gather unsprayed apples as available. There's no need to wash the apples, because the biology on the fruits' surfaces is desired. Do *not* use rotten apples.

Step 2. Quarter the apples and place them in a crock or jar so that the container is half to two-thirds full. Any container scale will be fine.

Step 3. Add enough good water to cover the apples. If desired, place a plate on top of the apples and position a jar filled with water on the plate as a weight to keep the apples submerged. The plate should be clean to start with to avoid contaminating the vinegar. Another option is to use a flat rock. Boil the rock first and allow it to cool before putting it in contact with the apples. The purpose of the weight is to prevent apples from floating to the surface of the liquid. Floaters can be removed if desired. Using a weight is also optional; fermentation will occur even with total neglect. I have found that even if I don't use a cover, a mother will form over time. The mother is that gelatinous SCOBY (symbiotic culture of bacteria and yeast) that forms on the surface of some fermentations.

Step 4. Place a cloth over the top the container and a board over the cloth to keep out rainwater or bugs. Label the container with the contents and the date. If you haven't done so already, position the container out of the sun in a place where it will not need to be moved for a few months.

Step 5. Soon the apples will begin to ferment and give off a wonderful smell. After about a month the fermentation will change. The pH will have fallen to about 4 to 5 at this point and may be decanted, but the mother will not yet have formed on the surface. The resulting liquid will be cloudy and not the reddish color of apple cider vinegar. If a weight is in place, remove it at this point.

Step 1

Step 2

Step 3

Step 6

Step 8

Step 6. The fermentation may be left for another month or longer, and a mother should form on the surface of the liquid in the container during that time.

Step 7. Use pH paper to monitor the pH of the liquid under the mother if desired. Simply dip the paper strip into the liquid and compare the color of the strip to the indicator chart on the pH paper package.

Step 8. When time permits, or when remembered, strain the liquid into glass jars and cover tightly for storage. Be sure to label the container with the date, length of fermentation, and ingredients. The apples may be added to the compost pile, completing a sustainable cycle.

Uses

You can use your homemade vinegar for any recipe in this book that calls for apple cider vinegar. Making your own further reduces your out-of-pocket cost as well as the waste and transportation costs associated with commercial-scale production of apple cider vinegar. Using locally available apples that might otherwise be left to rot on the ground and the rainwater that will flow into the earth anyway is the essence of a sustainable process. This may seem like a small thing, but is part of a larger philosophical approach for all agriculture processes and a main tenet of this book.

RECIPE AT A GLANCE

- Gather apples.
- Quarter apples and fill bucket three-quarters full; add enough rainwater to cover.
- Put weight on top to keep apples submerged; leave weight in place about 1 month (optional).
- Watch for mother to form.
- After mother forms or when pH drops below 5.0, decant through strainer.

Vinegar Extractions

Amendment type: Shelf-stable; mineral
Raw materials: Shells or bones that would otherwise end up in the garbage
Content: Broad-spectrum minerals and other compounds

Vinegar extractions are a great example of a sustainable agriculture amendment, because you can use apple cider vinegar you make yourself (see page 125) and locally sourced materials such as eggshells and animals bones left over from your home-cooked meals or gathered from a restaurant, butcher, or fisher on the docks if you live near the ocean. It's a good exercise in becoming aware of the free resources available in your area—be creative as you search.

The acidity of organic apple cider vinegar breaks down mineral constituents of materials into water-based forms that are shelf-stable for years. Vinegar extractions may be diluted and applied directly to plant roots, leaves, and bark as well as the soil.

One nice aspect of this recipe is that if the contents are forgotten and left to sit for months, that's okay! My kind of time constraint.

These bones, oyster shells, and eggshells have been cooked to prepare them for mineral extraction.

You can try this process with many types of animal products, and it's a great way to close waste gaps while making sustainable mineral amendments. Rather than throwing materials away, extract their minerals and feed those valuable mineral nutrients to plants and soil biology. Good choices include shells—egg, oyster, crab, or clam—and bones, including cow, fish, and pig bones. Any local restaurant that serves oysters probably throws out those shells each night. Chat with a member of the staff—ask whether you can leave a bucket on Friday afternoon for the kitchen staff to toss the shells into and pick it up Sunday morning. The bucket will contain extraneous stuff besides the oyster shells, but you can separate out the shells, wash them, cook them, and store them for future use. Crab shells, clamshells, eggshells, and other resources may be gathered in a similar way. And when you make a nutritious bone broth, save those bones for vinegar extraction, too. This process may be scaled to any level.

You can also extract minerals from the residual materials of fermented plant juice (see "Fermented Plant Juice" on page 134) using vinegar, maximizing the opportunity to collect the unique mineral concentrations of the plant materials.

Materials

Eggshells, bones, shells, or other available material
Oven, grill, or fire
Glass jar or crock

Organic apple cider vinegar
Piece of clean cloth
Strainer

Instructions

Step 1. Be sure to give thanks for the materials provided, gathered, and used. Then begin by cooking the materials to remove water and residual organic material. Eggshells can be toasted in a toaster oven—just pop them in and push the toast button one to three times. Be sure the toaster is set for light toast. Bones used for making a broth can be cooked on a grill afterward or in a large metal can over an open fire to remove moisture and any residual organic material. On a gas grill set at medium heat (300 to 350°F/150–175°C), bones will cook sufficiently in 40 to 60 minutes, depending on their initial state. To cook bones over an open fire, first punch several holes in the can to provide ventilation. Once the fire has burned down, nestle the can filled with bones into the bed of coals for the night. By morning moisture and residual organic material will be gone. Cooking bones outside allows any unwanted smells to drift away rather than stink up the kitchen. Oyster or clam shells may be cooked in the same way as bones; an indoor oven set at low heat will also work. It may work best to cook lightweight shells like crab shells in the toaster oven, as you would eggshells. It is most efficient to cook material in bulk and store the cooked material on the shelf, and then extract them as needed.

Step 2. When you want to make an extraction, add cooked materials to a jar or crock, filling about 10 to 15 percent of the container.

Step 3. Add organic apple cider vinegar, filling the container close to full. Leave at least 1 inch (2.5 cm) of space at the top, because fizzing and foaming may occur as chemical reactions start to take place. It's a good idea to put the jar or crock in a bowl or on a tray to catch spills.

Step 4. Put the cloth over the top of the jar or crock and store in a well-ventilated space out of sunlight for at least a week. Do not cover the jar or crock tightly, because the gases of reaction need to escape. A lid can be used to cover a glass jar during this phase, but leave it loose.

Step 4

Step 6

Step 5. After a couple of weeks, the extraction will have run its course, and the liquid may be decanted. Repeat extractions after decanting the liquid. There is still much mineral left in the bones or shells after the first extraction, so repeat the process until no gas reaction occurs. This may be three to five times.

Step 6. Strain and store the liquid in a glass jar with a secured top in a well-ventilated space out of sunlight. Be sure to label the container with dates, the extraction process used, and the ingredients. If you do multiple extractions of a batch of material, you can store all the decanted liquid in one glass container.

Step 7. Once you have extracted the bones or shells as many times as you wish, put the residual material into the compost pile. These cooked materials lack the odors that may attract most types of animals, although I have found leftover chunks of bones in trees around my property, gnawed on by squirrels that recognized the bit of mineral content that remained in them. Now, *that* is completing the waste cycle.

Uses

Shells are good sources for calcium; bones provide calcium and phosphorus as well. Both are rich in macro- and microminerals wanted in the soil and by plants.

These products may be used to augment the mineral content of broad-spectrum amendment products, such as fermented plant juice made from dandelions, to facilitate the mineral needs of plants at different growth phases. Dilute vinegar extractions with good water at a ratio of 1:500 up to 1:1000. (One tablespoon in 4 gallons / 15 L is approximately 1:500.) Apply as a foliar spray or a drench. Examples of the minerals in the liquid strained from vinegar extractions are provided in appendix E.

RECIPE AT A GLANCE

▶ Consider life history of materials used (pesticides, GMOs) before deciding whether to extract minerals from them.
▶ Give thanks for materials provided.
▶ Remove water and organic residue by cooking at 300–350°F (150–175°C).
▶ Mix in glass jar or crock. 10–15 percent material, 85–90 percent organic apple cider vinegar, by volume.
▶ Decant after a week or two and repeat until reaction no longer occurs.
▶ Remains go to compost pile.
▶ Strain and store out of sun in well-ventilated, constant-temperature space.
▶ Shelf-stable for years.
▶ Dilute 1:500 and apply as foliar spray or drench.

FERMENTED PLANT JUICE

Amendment type: A shelf-stable source of nutrients
Raw materials: Organic brown sugar, plants
Content: Broad spectrum of minerals and secondary
  compounds

Fermented plant juices are easily made, shelf-stable, low-cost, sustainable, and regenerative mineral amendments. These products take advantage of the great mineral accumulators in our backyard—the weeds and those less-than-perfect end-of-the-season fruits. These amendments also contain other beneficial compounds besides minerals: enzymes and proteins, all in forms plants can use.

Mineral-rich weeds like dandelion and stinging nettle are good plants to try first, because they are easily found and their broad-spectrum mineral proportions and amounts provide a good food source for the healthy growth and development of many kinds of plants. Experiment making FPJ with other plants, recognizing that each plant type has unique characteristics that may be harnessed. Use these in conjunction with other amendments to facilitate specific plant growth periods: germination, vegetative growth, reproduction, senescence. (Mineral needs of specific growth periods are discussed in detail in chapter 2.) Examples of the minerals measured in the liquid strained from many types of fermented plant juices are provided in appendix E.

You can scale up this recipe if you have a larger container. It will take about 2 pounds (1 kg) of plant material to fill a 1-gallon (4 L) crock, for example. The ratio of plant material to sugar ranges from 1:0.5 to 1:1. The more moisture in the fermenting material, the more sugar is required. Fruits require more sugar than leaves, for instance. This recipe can be scaled to any volume.

I first learned how to ferment plants in this manner by experimenting with the technique described in *Natural Farming Agriculture Materials* by Cho Ju-Young.

Materials

2-quart (2 L) crock or glass jar

About ½ pound (225 g)
   plant material

About ½ pound (225 g)
   organic brown sugar

1 quart (1 L) glass jar

Rock or glass of water

Scale

Sieve

Funnel

Clean dishcloth or equivalent

Instructions

Step 1. Pick plant leaves and stems early in the morning before the sun is up, while the dew is still on them. This is when the plant leaves have the most energy in them. During the day there is significant energy flow through the leaves as a result of transpiration and photosynthesis. At the end of the day and during the night, the plant stores energy within. By picking the leaves in the morning before the sun rises, this energy is captured for making fermented plant juice. Lactic acid bacteria and yeasts are captured on the plant surfaces. Do not wash the plants and do not harvest just after rain (because rainfall can also wash off lactic acid bacteria and yeasts). If conditions have been very dry for several days, it is okay to water the plants the day before you plan to gather material from them.

Step 2. Weigh the plant material and add organic brown sugar. For relatively dry plants, start with a ratio by weight of 3 parts plant material and 2 parts organic brown sugar. For wetter ingredients like fruit, use a weight ration of 1:1.

Step 3. Mix the plant material with the organic brown sugar in a large bowl. After mixing move the mixture into the glass jar or crock. Or place the unmixed plant material and sugar into the glass jar or crock in alternating layers. A tall, narrow container will work better than a wide, shallow one. Cover the top surface of the material with a layer of sugar. Initiate fermentation by placing the rock or glass of water on top of the contents in the jar or crock. Any item that will fit into the opening of the crock or glass jar can be used as a weight, but avoid items made of plastic, stainless steel,

Step 2

Step 3

Step 5

Step 6

Step 9

or other metals. The weight of the added item will press the sugar and plant material together, which will initiate osmosis. Be sure the weight is clean. Boil the rock before use, for instance.

Step 4. Cover the entire jar with the cloth and store in a dry, well-ventilated area, at room temperature (about 70°F/20°C) and out of sunlight.

Step 5. Periodically check on the jar. Within hours, a brown liquid will begin to form at the bottom. When enough liquid has formed so that all of the plant material is submerged, the weight may be removed. This may take a day or so. Allow fermentation to proceed for several days, continuing to monitor the jar to be sure the plant material stays submerged. After about a week, the fermentation process will be complete. The contents will have a fragrance that is unique and not off-putting. I have left fermenting plant material sitting longer than this and still gotten a stable product, but eventually the contents in the jar will change into something else that may not be as useful. This is something to experiment with, either intentionally or by default!

Step 6. Place the funnel into a glass jar and put the sieve atop the funnel. To strain the liquid from the plant matter, invert the fermentation container into the sieve so that the liquid drains freely into the quart (1 L) jar. This setup provides sturdy support for a glass jar ½ gallon (2 L) in size or smaller. For a crock or other heavy container, a different setup would be appropriate. Some imagination may be required.

Step 7. Let the mixture drain for several hours, overnight perhaps. Do not attempt to squeeze residual liquid out of the plant material. The amount of liquid amendment produced will vary depending on how much moisture was in the plant matter at the start of the process. If foam appears on the top of the fermentation liquid after extraction, add more sugar to eliminate it. Take note of these details and record them in your garden notebook for future reference.

Step 8. The residual material can be further extracted to produce shelf-stable, mineral-rich amendments using organic apple cider vinegar as described in the "Vinegar Extractions" recipe on page 129. Vinegar extractions of the residual plant matter can be carried out several times. In general, the remaining mineral

constituents in the plant residues will be less than that captured in
the fermented plant juice, but sometimes I have obtained higher
concentrations of specific microminerals in vinegar extractions
of residues. This is an example of how to obtain higher concentrations of specific microminerals in an amendment, based on an
informed choice of plant type.

Step 9. Label containers with the plant type and parts used and the
dates when fermentation was started and ended. Add to your
garden notebook the weights of plant and sugar used, the type
of sugar, the time of day the plant was harvested, the containers
used, and other information deemed appropriate. Store in a
well-ventilated, constant-temperature area out of sunlight.

Uses

Using fermented plant juice amendments offers a balanced approach
to plant nutrition. I suggest using single-plant fermentations to start
with. Take good notes on what you apply and the kind of results you
see. This is an amazing product to learn about and experiment with.
Fermented plant juices can form the basis of your short-term mineralization efforts throughout the growing season.

Fermented plant juices contain nutrients in plant-available forms.
Apply them directly to plant foliage and to bark if no foliage is available
(think of trees in late fall through early spring) as a foliar spray, or apply
as a drench to feed plant roots and soil biology. The optimal dilution
ratio will depend on the type of plant fermented and the needs of the
plant being fed. Start with dilution ratios of 1:1000 or 1:500. (A homeopathic approach is something to consider, as discussed in chapter 5 on
page 115.) Diluting fermented plant juices in a 5-gallon bucket filled
with about 4 gallons (16 L) of good water works well, because there's
enough space at the top to allow for vigorous stirring. One tablespoon
of fermented plant juice added to 4 gallons of water is a 1:1000 ratio, 2
tablespoons a 1:500 ratio. If a quart of FPJ is made in the kitchen and
diluted 1:500, this will make 500 quarts (500 L) of amendment—that's
125 gallons! Small amounts of this product go a very long way.

Fermented plant juice is also valuable for encouraging germination.
Seeds may be soaked prior to planting. Dilute the fermented plant juice

in the same ratio as above and soak seeds for several minutes prior to planting so that the minerals needed for growth are available as soon as sprouting occurs. The continued exposure to fermented plant juice will assure that minerals are availabe in forms the young seedlings can access, reducing potential sources of stress as a result of poor nutririon.

RECIPE AT A GLANCE
▸ Pick plant material before sunrise when it is wet with dew; do not wash material.
▸ Mix plant material with organic brown sugar.
▸ Fill crock or glass jar; cover top surface of material with sugar.
▸ Initiate fermentation by putting weight on top of contents. Cover it with cloth.
▸ Allow liquid to form until material is submerged, then remove weight.
▸ Store out of sunlight at room temperature for 1 week.
▸ Strain liquid, using force of gravity only.
▸ Label container.
▸ Store at room temperature, well ventilated, out of sunlight.
▸ Use as foliar spray or drench, diluting with good water starting at 1:500.

Fermented Fish

> **Amendment type:** A shelf-stable mineral amendment
> **Raw materials:** Fish waste from local market or dock (or a fishing trip), organic brown sugar
> **Content:** Broad-spectrum minerals and other compounds

This garden amendment makes use of nutrient-rich material that often goes to waste in a fish market, at a grocery store or restaurant, on the docks of a fishing port, or at other local places. When you set out to gather fish wastes for this recipe, keep in mind that we are what we eat and fish are no exception. Wild-caught fish are preferable to farm-raised fish that may have been fed genetically modified grains or other foods that are not what nature intended. Bottom feeders such as catfish eat from areas where heavy metal toxins may accumulate. Young, oily fish like bluefish, mackerel, or anchovies caught from the open ocean are a top choice if you can get them. Freshwater fish do not have the benefit of the minerals in the oceans, but still contain minerals and compounds useful to the soil and plants. Remember, the Indians used fish to feed the three sisters: corn, beans, and squash.

All parts of a fish are valuable for making amendments. The parts generally discarded after fish are filleted are great options to complete a sustainability cycle rather than to fill up landfills. Use what is available especially if it will close the waste gap. Catch your own fish. What a great reason for a day at the beach! Vinegar extraction of the material left after fermentation of fish further breaks down the material into valuable water-soluble amendments.

Materials

Approximately 8 pounds (3.5 kg) fish (give thanks to the fish)
Approximately 8 pounds (3.5 kg) organic brown sugar
2-gallon (8 L) crock
Clean piece of cloth
Strainer
Approximately 7 quarts (7 L) organic apple cider vinegar
8 quart-sized canning jars with lids, or other sealable container of equivalent volume

Instructions

Step 1. Give thanks to all life used to nurture soil and plants, both animal and plant life.

Step 2. Weigh the fish. If it's not precisely 8 pounds, that's okay, but make a note of the poundage. Whatever the weight, you will combine it with an equal quantity of organic brown sugar (1:1 ratio by weight).

Step 3. Chop the fish into chunks and mix with the equal weight of organic brown sugar in a large bowl.

Step 4. Put this mixture into the crock. Be sure there is adequate air space above the mixture—the crock should be two-thirds full. If there is more fish than will fit, get a bigger crock, fill a second crock, freeze the fish for later use, or layer the excess in the compost pile.

Step 5. Cover the top of the mixture with a layer of organic brown sugar. If IMO #4 is available, sprinkle some of it over the sugar. (Making IMO #4 is a multistage process; see the IMO recipes beginning on page 158.) The biology in IMO #4 consumes the anaerobic constituents on the surface and will virtually eliminate any unpleasant smells during the fermentation process. I once fermented 64 small bluefish in this manner in my living room for about 6 months, with no complaints from the odoriferously astute members of the household, if you know what I mean!

Step 6. Cover with a cloth and allow to ferment for about 6 months in a dry, well-ventilated place at room temperature (about 70°F/20°C) and out of sunlight. I like to monitor the appearance of the surface of the material and the smell of the crock over time, but this is not necessary. After about 6 months, the contents below the surface will be liquid and ready for straining. The crock may be heavy, so carefully strain the contents into a large bowl and then transfer the liquid in the bowl into a glass container with a lid. Store this liquid in a dry, constant-temperature, well-ventilated space out of sunlight. Be sure to label the container with the type of fish, the source, and the dates of the beginning and end of the fermentation cycle. Add to your garden notebook the quantities of fish and sugar, along with any other pertinent processing information.

Step 7. In order to capture as much nutrient as possible from this valuable resource, I recommend further extracting minerals from the remaining solids using organic apple cider vinegar. (See "Vinegar Extractions" on page 129.) Multiple vinegar extractions can reduce the residual solids until nearly nothing is left. Four or more extractions may be completed. Whatever remains can be added to a compost pile. No waste from this valuable gift.

Uses

Apply these products whenever broad-spectrum nutrition is wanted. Dilute 1:1000 with good water and use as a foliar spray or drench. It can also be valuable when making IMO #3 and IMO #4. One quart of fermented fish diluted 1:1000 will make 1,000 quarts of amendment—that's 250 gallons (almost 1,000 L). Small amounts of this product go a very long way.

If you carry out vinegar extractions with the remains from the first ferment, keep in mind that each extraction will contain slightly different spectra of minerals. Examples of the minerals measured in the liquid strained from a fermented fish extraction and subsequent vinegar extractions of the residue from fermentation are provided in appendix E.

RECIPE AT A GLANCE

- Use available fish, closing waste gap—oily, wild-caught ocean fish are best.
- Give thanks to fish that will feed soil, plants, and you.
- Chop and mix with organic brown sugar 1:1 by weight.
- Fill crock two-thirds full with mixture.
- Cover material with top layer of organic brown sugar and IMO #4 if available; cover with cloth.
- Ferment in dry, well-ventilated space out of sunlight for about 6 months.
- Strain liquid through sieve and store in clean glass jar with lid.

- Label with dates, process information, and type of fish used.
- Use organic apple cider vinegar to extract minerals from fermentation residue.
- Store all final products in dry, well-ventilated space out of sunlight.
- Dilute with good water 1:1000.
- Use as foliar spray or drench.

LEAF MOLD FERMENTATION

Amendment type: A shelf-stable mineral and biological amendment

Raw materials: Rainwater, leaf mold, plants

Content: Broad-spectrum minerals, biology, and secondary compounds

Leaf mold is the layer of decomposed leaves beneath the surface of the woodland floor, just under the thin top layer of dry leaves that have not yet decomposed. Leaf mold is a rich source of soil biology. Using this biology source to ferment plants in water significantly reduces the unpleasant smells that would otherwise develop during fermentation using just water and plant matter.

Plant matter mixed with water and leaf mold will break down over time to produce a customized amendment based on the unique

Leaf mold contains thousands of species of bacteria, fungi, and archaea that thrive in the local environment.

mineral composition of the plant. You can scale this recipe to any level required. If growing acres of a single crop, use the end-of-the-season crop residue (leaf, stem, and/or fruit) to produce a cheap and extremely effective amendment for next year's crop. At a dilution ratio of 1:20, even 5 gallons of leaf mold fermented material goes a long way as a foliar spray (20 × 5 = 100 gallons). A 50-gallon container goes even farther (20 × 50 = 1,000 gallons)! Remember that Dr. James Duke's Phytochemical and Ethnobotanical database is a great resource for learning about the unique mineral proportions of individual plant species. (See "Weeds and Crop Plants" on page 113 for more about this database.)

Examples of the minerals measured in the liquid strained from several leaf mold fermentations are provided in appendix E. I adapted this recipe from information in *JADAM Organic Farming* the way by Youngsang Cho.

Materials

Bucket

Plant residue (leaves, fruit, stems) to fill bucket ¾ full

Good water

Leaf mold (gathered from a local wooded area)

Wooden board

Instructions

Step 1. Gather leaves or fruits of a specific plant type. Fill the bucket three-quarters full with plant material. Cut material up to increase surface area.

Step 2. Add enough water to submerge the contents of the bucket, along with a handful of leaf mold.

Step 3. If desired, a plate with a weight could be placed on the top to submerge the contents.

Step 4. Cover the bucket with the board to keep out rainwater and debris, and store out of the sun in a well-ventilated space. Bring the bucket inside during the winter to prevent freezing if required.

Step 5. Allow the bucket to sit and ferment—this is a "garden time" amendment that will take months to mature. When I make this amendment, I generally plan to store the bucket overwinter for use the following year.

Step 6. The longer you allow the fermentation process to proceed, the greater the amount of mineral content will be incorporated into the liquid solution. As the ferment proceeds, you can strain off some of the liquid from the bucket to apply to your crops, and add more water to refill the bucket. More plant residue may be added.

Fermenting Peaches

I once made a leaf mold fermentation using small, bruised, and fallen peaches from my backyard peach trees. To start with, I simply put the peaches into a crock and covered them with a cloth. A month or two later, the peaches were in a liquid state. I stirred the contents to break apart the remaining intact peaches, spread a thick layer of leaf mold across the top, and covered the crock with a cloth and then a board. Months later, as cold weather threatened to freeze the crock, I took off the cover and found a thick mother atop the liquid. I could have moved the crock indoors and let the decomposition process continue, but instead I strained the contents, adding the nutrient-rich liquid to my shelf of amendments so I could use it to nurture my peach trees the following year. I spread the remaining fruit material at the base of my peach trees.

I have used this method of spreading a thick layer of leaf mold on the surface of other fruit fermentations and found the contents absent of any undesirable smells as fermentation progressed. This forgiving process is the ultimate low-cost, sustainable, regenerative way to make custom garden amendments.

Uses

The liquid produced by leaf mold fermentation will contain a broad spectrum of minerals as well as enzymes and other complex compounds. The precise content will reflect the unique characteristics of the plant material used to make the amendment, which is motivation to learn about the weeds in your area and their mineral content.

Dilute the liquid with good water in a ratio of 1:20 or more and apply as a foliar spray or drench. Experiment with different dilution rates for optimal results.

RECIPE AT A GLANCE

- Fill bucket ¾ full with plant residue.
- Fill up bucket with good water.
- Add handful of leaf mold or cover liquid surface with leaf mold.
- Cover with cloth and board to keep out rain and bugs.
- Ferment for up to a year; longer increases mineral constituents.
- Store out of sun or rain. Prevent from freezing.
- Can be strained and resulting liquid stored on shelf.
- Dilute 1:20 or more and use as foliar spray or drench.

Leaf Mold Biology

Amendment type: Short-lived biological amendment
Raw materials: One potato, leaf mold
Content: Local biological diversity

Leaf mold is the quintessential local source of biology; and leaf mold biology is the very definition of a sustainable, regenerative amendment. It is easily made for nearly no cost and represents the most diverse local and seasonal biological inoculants available: bacteria, fungi, and archaea.

Leaf mold gathered near gardens or farm fields contains thousands of types of bacteria, fungi, and archaea that thrive in the local environmental conditions. The range of species present in leaf mold is beyond our current understanding. Thus, this amendment provides an influx of diverse life that will transform thick, heavily mulched soil into airy, tilthy soil, facilitate compost digestion, or inoculate soil with a thriving biological ecosystem for seeding or transplanting. In this recipe a potato supplies starches to feed the organisms present in the leaf mold, resulting in a bucket filled with local life. As discussed in part 1, in order for plants to thrive, the soil biology must thrive. Leaf mold biology will significantly and quickly improve soil structure by introducing local biology, digesting organic material, and jump-starting the soil ecology.

Treat this amendment with love: It is a living product that needs housing, food (organic carbon), and water to survive. Mulch, compost, and cover crops can meet these needs! Adjusting our understanding and gardening practices to fulfill the need for organic carbon to feed the soil biology is one of the most significant shifts involved in becoming a sustainable grower.

Leaf mold biology can be made throughout the growing season whenever temperatures are above 40°F (4°C). Making and applying leaf mold biology to growing areas periodically facilitates the seasonal variation in the makeup of soil biology. Consider applications between plantings, whenever mulch is applied, on new and fallow garden plots, and at the beginning and end of the season.

I adapted this recipe from *JADAM Organic Farming* by Youngsang Cho, and it may be scaled to any level. Variations depending on available materials is fine. A burlap bag may take the place of the sock, for instance.

Materials

5-gallon bucket
Good water
About 1 tablespoon sea salt
1 medium-sized boiled potato
Handful leaf mold
2 white cotton socks

2 small rocks
2-foot (60 cm) length of
    flexible wire or string
1 paper clip
Wooden board (large enough to
    cover top of bucket)

Instructions

Step 1. Gather leaf mold from just below the surface layer of undecomposed leaves on the floor of a woodland near the crop growing area.

Step 2. Assemble all materials. Boil the potato until it softens. The rocks should be just large enough to weigh down the socks so they remain suspended in the water in the bucket—a rock that weighs about 1 ounce (28 g) should work well.

Step 3. Twist one end of the wire or string around one end of the bucket handle and stretch it across the mouth of the bucket. Secure the other end around the other end of the bucket handle.

Step 4. Place one of the rocks and the boiled potato in one sock. In the other sock, place the other rock and the handful of leaf mold.

Step 5. Add good water to the bucket, filling it up to 2 to 3 inches (5–7.5 cm) below the rim. Add the sea salt to the water and stir until dissolved.

Step 6. Open up the paper clip and split it in half. Use the clip pieces to support each sock on the wire so that its contents hang down into the water between the surface and the bottom of the bucket at a height approximately in the middle of the bucket.

Step 7. Knead and squeeze each of the socks to force much of the contents through the fabric and into suspension in the water. Continue until the potato is well smushed and the water has turned light brown in color from the leaf mold. This will take only a couple of minutes. This action spreads the biology from the leaf mold and the starch from the potato (which will feed the biology)

throughout the water in the bucket. The rocks will keep the contents remaining within the socks submerged in the water.

Step 8. Leaving the socks in place, position the bucket outdoors near the spot where the amendment will be applied. Place the wooden board over the mouth of the bucket as a cover to keep out rain and debris.

Step 9. Check the contents of the bucket daily. After 1 to 5 days, foam will appear on the surface of the water. In warm summer weather, foam may form in 1 day. On cooler days with temperatures from 45 to 55°F (7–13°C), this may take several days. The thickness or robustness of the foam will vary with air temperature as well. Warmer temperatures will produce a thicker foam, cooler days a thinner one. Foam thickness will increase and then decrease, representing the flourishing of the biology in the solution and its subsequent decline. When the foam seems to have reached its maximum, the amendment is ready for use as a biological stimulant. A thin space of no foam will appear at the edge of the bucket.

Step 10. Remove the socks from the bucket. Rinse the socks and rocks in clean water for use next time around.

Applying a wide spectrum of biology causes competition for available resources, which causes plant and soil pathogens to stay in check.

Uses

When the foam is at its peak, stir the contents of the bucket and dilute 1:10 or up to 1:50 with good water. The biology will be suspended throughout the liquid, not just on the surface. The surface foam is merely an indication of the ripeness of the contents. This product may be applied as a foliar spray or a drench.

Experiment with leaf mold biology and learn how powerful this free resource is on the farm and in the garden. Here are some suggested uses for the diluted leaf mold biology liquid; I am sure there are others to be discovered:

- Apply to heavily mulched areas where the mulch has become dense and unable to absorb water. If a mulched area has developed a particularly hard surface, using a tool to disturb the surface can provide penetration sites, but in general, this is not necessary. The biology will quickly digest the organic carbon, and within days the area will be aerated and able to absorb water, lots of it. I have had to mulch areas two and three times per year because the biology so quickly digests the organic carbon mulch applied.

- Apply to compost piles to stimulate biological activity.
- Apply as a drench at the beginning of the growing season to stimulate soil function.
- Apply as a drench at the end of the growing season to stimulate soil function.
- Apply as a foliar spray to thwart biological pathogens on leaf surfaces.

After the foam on the top of the bucket has cleared up, the contents no longer contain a biological source, but the liquid can still be used as an effective mineral amendment. This may be decanted through a strainer or simply applied. Dilute 1:10 or up to 1:50 with good water and apply as a drench or foliar application.

RECIPE AT A GLANCE

- ▸ Affix wire or string across top of bucket, fastened at both ends to bucket handle.
- ▸ Put boiled potato and rock into sock.
- ▸ Put handful of leaf mold and rock into second sock.
- ▸ Fill bucket with good water, stir in sea salt.
- ▸ Suspend and secure socks over wire so sock contents are immersed in water.
- ▸ Knead socks so that contents are distributed into water.
- ▸ Cover bucket and place it in environment where amendment will be used.
- ▸ Monitor; after 1 to 5 days, foam will appear on surface.
- ▸ When quantity of foam peaks, use immediately as biological amendment.
- ▸ Dilute 1:10 or up to 1:50 and apply to soil or mulch to stimulate soil life.
- ▸ After foam has dispersed, remaining liquid can be used as mineral amendment.

Lactic Acid Bacteria

Amendment type: A shelf-stable (when refrigerated)
  biological amendment
Raw materials: Water from rinsing rice, raw milk
Content: Broad-spectrum lactic acid bacteria

Lactic acid bacteria (LAB) is a refrigerator-stable biological product. Having a stable biology source available year-round is very useful, especially in early spring when the soil biology has not yet come to life.

Lactic acid bacteria are a group of acid-tolerant bacteria that produce lactic acid as a metabolic end product of carbohydrate fermentation. The acidification inhibits the growth of other pathogenic microorganisms, and thus these bacteria are often used for the fermentation of foods. (Think sauerkraut, pickles, and the like.) It is preferable to use raw milk when making this amendment because raw milk contains additional lactic acids and other nutritional compounds that may be killed or altered by the process of pasteurization and homogenization. Still, any milk will do, so use what you have.

This method is adapted from a recipe in *Natural Farming Agriculture Materials* by Cho Ju-Young.

Materials

2–3 cups (370–550 g)
  organic rice
Sieve
1 quart (1 L) good water
1 quart-sized canning jar with lid

1 large bowl
Raw milk
1 pint-sized canning jar
  with lid
Clean piece of cloth

Instructions

Step 1. As you prepare organic rice for cooking, or for use to capture IMO #1 (see page 158), rinse the rice through a sieve with a quart of good water in the normal way, but collect the rinse water in a bowl set under the sieve. Put the collected water back into the jar and use it to rinse the rice again. Repeat this rinsing step several times.

Step 2. Pour the rinse water into the quart-sized canning jar and cover with the cloth. Label the jar with the date and contents.

Step 3. Let sit for 3 to 5 days in a spot out of direct sunlight, at room temperature, with good air circulation, perhaps on your kitchen counter. During this time, the contents will separate into three distinct layers: a thin top layer, a clear middle layer, and a layer of residue at the bottom. The top and bottom layers may be slight and difficult to discern.

Step 4. The clear middle layer contains a high concentration of lactic acid bacteria. Extract the clear middle layer by first spooning off the top layer. Then pour the contents of the jar into another jar, leaving the bottom layer in the original jar. The residual materials (top and bottom layers) may be put onto the compost pile. Label the filled jar "Lactic Acid Bacteria Pure Stock" and include the date. This pure stock can be stored in the refrigerator long-term (years) and used as needed to make the lactic acid bacteria amendment, which can be diluted and applied to plants and soil.

Step 5. When you want to make the lactic acid bacteria amendment, mix pure stock with milk at a ratio of 1:10.

Step 6. Cover the container with a cloth and store for 5 to 7 days out of the sun, at room temperature, in an area with good air circulation, perhaps on your kitchen counter. As it sits, the contents will divide into three distinct layers: a thick cheeselike layer on the top, a yellowish clear layer in the middle, and a layer of sediment at the bottom of the jar.

Step 7. Extract the yellowish clear layer in the middle using the same method described in step 4. This liquid is lactic acid bacteria. Be sure to label the jar; it can be stored in your refrigerator for years. The cheese and sediment can be fed to chickens or any other pets or livestock that seem to like eating it.

Uses

This amendment is an excellent biology source for invigorating soil digestion and improving soil tilth. In *Natural Farming Agriculture Materials*, Cho Ju-Young suggests the following uses of LAB at a dilution of 1:1000 with good water.

Step 1

Step 4

Step 5

Step 6

Step 7

- Improving soil ventilation.
- Cultivating IMO #3.
- Accelerating root growth at transplant.
- Use during vegetative growth period.
- Solubilizing phosphate.
- Improving digestive function of animals.

Here are several other uses of LAB:

- Apply LAB to a compost pile to stimulate biological activity.
- Add LAB to the soil with IMO #4 to further transform stodgy, compacted soil into loose, fluffy soil.
- Apply LAB to the soil a week or two prior to planting seeds.
- Thoroughly soak the root ball of plants with LAB and other mineral amendment products prior to transplant.
- Use in conjunction with mineral amendments as a foliar spray or drench.
- Use as a foliar spray on leaves to combat airborne pathogens.

Applying a wide spectrum of biology causes competition for available resources, which causes plant and soil pathogens to stay in check.

RECIPE AT A GLANCE

- ▶ Rinse organic rice through sieve with good water.
- ▶ Collect rinse water, pour into jar, label.
- ▶ Cover jar with cloth and let sit 3 to 5 days out of sun with good air circulation.
- ▶ Extract middle layer of liquid (pure stock) from jar.
- ▶ Pour into separate jar, label, and store in refrigerator.
- ▶ Mix pure stock with raw milk 1:10.
- ▶ Cover container with cloth 5 to 7 days out of sun with good air circulation.
- ▶ Extract middle layer of liquid.
- ▶ Pour into separate jar, label, and store in refrigerator.
- ▶ Dilute 1:1000 with good water for use as drench or foliar spray.

IMO #1: Capturing
Local Microorganisms

Amendment type: A short-lived biological amendment
Raw materials: Wood, screws, and wire to make trap; rice
Content: Diverse local bacteria, fungi, and archaea

The new plant model recognizes the importance of soil biology. Culturing indigenous microorganisms (IMO) is an exciting and effective method to capture soil microbes and introduce them into the growing area. Methods of harnessing microorganisms from the soil are documented in both Japan (Bokashi, effective microorganisms) and Korea (indigenous microorganisms). It would not be suprising to learn of other cultures practicing the same or similar processes. I first learned about making IMO from *Natural Farming Agriculture Materials* by Cho Ju-Young, and for consistency, I follow the nomenclature of IMO #1, IMO #2, IMO #3, and IMO #4 originated by Cho to identify the four separate processes used to make the finished product (IMO #4). A handful of leaf mold from the floor of the local woods represents the broadest range of bacteria, fungi, and archaea available. Capturing this most valuable resource is a great introduction to the diversity and useful characteristics of the biology within the soil. (An overview of the full four-stage process of making IMO is presented in "Indigenous Microorganisms" on page 48.)

The populations of soil microorganisms are unique to a site and are influenced by its drainage, orientation to the sun, and elevation. Microbial makeup also varies with the seasons. Microorganisms captured at a lower elevation may not be as effective if introduced into soils at a higher elevation, but those captured at a higher elevation will be effective when introduced at a lower elevation according to Cho Ju-Young in *Natural Farming Agricultural Materials*. This makes intuitive sense, and is analogous to survival of plants at different elevations. Hardy plants that grow at higher elevations, where weather conditions tend to be more extreme, generally can survive at lower elevations, where conditions are milder. But plants adapted to grow at lower elevations may not be able to withstand the colder temperatures, stronger winds, or lower oxygen content of the air up high.

The instructions I provide in this recipe and the three that follow will produce about half a yard of IMO #4 for a garden and home landscape. IMO #4 production can be scaled up to any level. In this first stage of the process IMO #1 is made, which is not shelf-stable and is primarily used to make IMO #2.

Make a trap box out of cedar or other wood that is available. The dimensions cited in the recipe are not the only possibility. Any box large enough to hold the cooked rice and still have ample open space at the top to allow the microorganisms room to flourish may be used. A ratio of about one-third air space to two-thirds rice is good. A basket wrapped up in a piece of cloth will work, but that setup is more vulnerable to invasion by critters. Don't get hung up on finding exactly the right box. The most important thing is to try the process and learn from successes and mistakes. A sturdy trap should serve well for making many batches of IMO over time.

It's a good idea to scout locations for setting the trap before starting the process. Look for a place in the woods where the leaf litter is thick. Or, if there are is no woodland nearby, choose a spot near a big old tree or near a compost pile.

Animals dug up this wooden box in which IMO #1 was being cultured, but the screened cover stayed in place and saved the IMO for use.

Materials

Wooden (cedar) box, 12 × 9 × 4 inches (30 × 23 × 10 cm)

3 cups (550 g) organic brown rice

Piece of wire mesh, approximately 18 × 15 inches (45 × 40 cm)

Clean cloth

Plastic sheet or tarp

4 screws

Screwdriver

Shovel

Instructions

Step 1. Test out the box by fitting the wire mesh over the box opening, folding the mesh over the sides and overlapping it at the corners. Screw the mesh to the box at the four corners to secure it. The mesh screen is not essential to the process, but it provides protection against raccoons or other foragers. Also check that the cloth is large enough to cover the top of the box and extend over all four sides. The cover prevents soils and other debris from contaminating the rice and the developing biology, but it should be easy to peek under the cover to see how things are going inside the box. When the trap, screen, and cover are working well together, remove the screen from the trap and go on to the next step.

Step 2. Rinse and cook the organic rice in the usual way. Keep in mind that the water used to rinse the rice before cooking can be used to make lactic acid bacteria! (See "Lactic Acid Bacteria" on page 154.)

Step 3. Allow the rice to cool, and scoop it into the box. Be sure to leave the top third of the box empty so that the indigenous microorganisms will have space to grow.

Step 4. Cover the box with the wire mesh and use screws to secure the mesh at the corners.

Step 5. Cover the wire mesh with the cloth. The trap is now ready to be buried in a biologically rich location.

Step 6. Dig a hole at your chosen spot big enough to bury the box, such that the top surface of the trap will be just below the ground surface. Digging a hole in the woods may be a new experience. Start by removing the leaf litter on the surface and placing it to the side. The composition of the soil beneath the litter may be significantly different than in a garden. Root concentrations may be thick. Be nice, and work carefully. Those roots were there before you, and they circulate information and nutrition among the living. Set the box (with the cloth cover over the top and draping over the sides) in the hole and cover it with the leaf litter.

Step 7. Use sticks to mark the location of the trap.

Step 8. Leave the box in the ground for 5 to 10 days. Be mindful of the weather during this time. If heavy rain occurs, cover the trap area with a tarp. Light rain should not be a problem. Remove the tarp when the rain stops so the area can breathe.

Step 9. Check the contents of the box after a few days to see how things are progressing. The speed of microbial development is temperature-dependent. In early spring when soil temperatures are cooler—about 50°F (10°C)—this may take 10 days, while during the warm days of summer, 5 days should be enough. To inspect the contents of the box, remove the leaf litter and pull back the cloth. If no fuzz has developed, cover the box, wait a couple more days, and check again.

Step 10. When fuzz does develop, it's important to assess the appearance. It is desirable to have the contents filled with a white fuzz that resembles cotton. These are the local indigenous

microorganisms. If white fuzz has started to grow on the surface of the rice but unfavorable weather is predicted, it's okay to bring the box inside. Warm indoor temperatures will accelerate growth, so be ready to make IMO #2 within a day or two, before the fuzz in the box turns gray.

Step 11. If the fuzz is gray, the microorganisms have grown past peak and are on the demise. If just a little gray on the surface of the fuzz, then move to make IMO #2 immediately, as it is dying and will not last long.

Step 12. If the fuzz is mostly gray or black, the biology is dead. Discard the contents onto the compost pile and try again. If the fuzz has a significant amount of red and blue coloring, it should also be added to the compost pile and the process should be restarted. A small amount of discoloration is okay.

Uses

The biology captured by this culturing process will not survive for long. It should be immediately fermented to make IMO #2, which is a refrigerator-stable biological product. I imagine that IMO #1 could be spread over the surface of a damp area outdoors and the biology might prosper, but I have not experimented with this approach.

RECIPE AT A GLANCE
▸ Cook rice. When cool, fill box about ⅔ full.
▸ Cover box with wire mesh; screw mesh in place at box corners.
▸ "Plant" box in biologically rich area such as leaf litter, compost pile.
▸ Cover with cloth, leaf mold, dry leaves.
▸ Leave for 5 to 10 days, depending on weather (cold takes longer).
▸ Cover with tarp if heavy rain.
▸ White fuzz is local biology.
▸ Small amount of discoloration is okay.
▸ Too much discoloration is problematic; add material to compost pile and try again.

IMO #2: Fermenting Local Microorganisms

Amendment type: Shelf-stable (refrigerated) biological amendment
Raw materials: IMO #1, organic brown sugar
Content: Diverse local bacteria, fungi, and archaea

IMO #2 is the second stage of the IMO process, and it involves fermenting IMO #1. The fermentation process puts the local microorganisms into a suspended state that can be stored in the refrigerator for over a year and used when needed.

Materials

Captured local microorganisms (IMO #1)
Approximately 3 pounds (1.4 kg) organic brown sugar
Scale
Large mixing bowl
1 half-gallon glass canning jar and 1 quart-sized glass canning jar
Piece of clean cloth
1 glass canning jar with tight-fitting lid for storage

Instructions

Step 1. Once white fuzz is established in your local microorganism trap (see "IMO #1" on page 158), remove the contents (both rice and white fuzz) from the box and weigh it. Add the same amount by weight of organic brown sugar.

Step 2. Mix the materials all together in the bowl.

Step 3. Transfer into the crock or glass jar for fermenting. Be sure to leave air space above the contents, because the fermentation process needs to breathe. A good ratio is two-thirds ferment, one-third air space. Assuming 3 cups of rice was used when the IMO #1 was made, you will need to split the rice-sugar mixture between a ½-gallon and quart jars for fermenting. Other options would be to get a larger jar or fill the ½-gallon jar two-thirds full and discard the rest onto the compost pile.

Step 4. Cover the contents with the cloth and allow to ferment in a well-ventilated area out of the sun at a constant temperature for 7 to 8 days. The contents will morph into a brown liquid with a unique smell that is not totally disagreeable.

Step 5. After fermentation, transfer the contents into the glass jar(s) and screw on the lid. Label the container with the date and the location from which the local biology was captured. Store in the refrigerator.

Uses

IMO #2 is refrigerator shelf-stable for several years. The primary use of IMO #2 is to make IMO #3, but I have also used it to inoculate soils, seeds, and compost piles, especially if no other inoculant is on hand. Dilute IMO #2 at a ratio of 1:1000 and apply to the soil as a light watering to initiate biological activity before planting or to stimulate spring soil. To be effective there must be water, housing, and food to support the added biology! Adding to bare dried soil in the middle of the day will not be as effective as adding it to lightly mulched loose soil at the end of a rain, for instance. Add to the compost pile in the same manner to stimulate biological activity there. I have also used IMO #2 as a seed inoculant by adding the smallest amount to a jar of good water with minerals and submerging the seeds into the solution for 10 to 15 minutes before planting.

This is a unique product with many uses still to be established. Recognize it as a biological resource and experiment with it. Use your imagination.

RECIPE AT A GLANCE
- Weigh IMO #1 (rice and white fuzz).
- Mix with equal weight of organic brown sugar.
- Cover and allow to ferment for 7 to 8 days.
- Put contents into glass jar, screw on lid, and label.
- Store in refrigerator.

IMO #3: PROPAGATING LOCAL MICROORGANISMS

Amendment type: A biological and mineral amendment
Raw materials: IMO #2, organic wheat bran, homemade mineral amendments, good water
Content: Diverse local bacteria, fungi, and archaea; other compounds, minerals

In this process a pile of organic wheat bran is inoculated with the local biology contained in IMO #2. This is similar to starting a compost pile in that biology must be established throughout before significant digestion can occur. Thus, this stage is a multiplication phase that will spread the biology through a much larger volume of material. I have batches of IMO #2 made at different times during the growing season, and I mix them together when inoculating wheat bran to make IMO #3. Making IMO #3 is a lot of work for one person to do alone, but it can be a wonderful task for a group of people, bringing together their energy for a positive result.

This phase is a good opportunity to add water-soluble mineral amendments, such as fermented plant juices, vinegar extractions, and leaf-mold-fermented juices. This is done by adding small amounts of these extractions to the water used to inoculate the wheat germ. The mixing will distribute the small amount of trace minerals throughout and establish desired trace mineral proportions within the IMO #4, the final product of this multistage process. Use soil test results as a guide to determine concentrations of minerals to add. The goal may be to remedy a specific known soil deficiency. Or it may be to establish a balanced mineral composition in the IMO #4. The mineral constituents of analyzed amendments listed in appendix E may be useful for determining which amendment products to add to the water when making IMO #3.

Organic wheat bran is available at any good-quality grain mill. The wheat bran is a by-product of the milling process; it typically costs about $60 for 150 pounds. This makes IMO #3 the most expensive amendment product I describe in this book. I have experimented with inoculating other types of materials but have not identified a cheaper local material yet. I am sure an alternative to wheat bran is out there, waiting for one of us to discover and share the news.

It's worth repeating: Be sure to take notes during this process. Write down the amount of ingredients added to the pile, the temperature of the pile, how many times the pile was turned to maintain temperature, when the bloom of microbial fuzz appears, and anything else that comes to mind. This will be a useful reference the next time around. This process is fairly forgiving, so don't get uptight about the details. Just give it a try, take notes, and see what happens.

In advance, scout out the spot where you will spread the bran and mix the pile. Look for an area that will be shaded for most of the day. The shade will reduce moisture loss, and the pile will stay cooler than it would if exposed to bright sun during the heat of the day. The nearby trees will benefit from the residual biology and minerals left at the site even after collection. If a suitable new site is used each time IMO #3 is made, it will help to spread these benefits around the landscape.

Materials

4–6 tablespoons IMO #2

150 pounds (67 kg) organic wheat bran

Approximately 35 gallons (128 L) good water

Buckets

Water-based mineral amendments (optional but recommended)

Shovel

Thermometer

Wheat bran wrapper bags, leaves, or straw

Instructions

Step 1. Remove all green matter from a chosen site that is mostly in the shade during the day, exposing bare ground. Pile the organic wheat bran on the bare ground.

Step 2. Add about 4 gallons (15 L) of good water to each bucket.

Step 3. Add 1 tablespoon of IMO #2 (1 tablespoon in 4 gallons is about 1:1000) to each bucket. It is not necessary to add IMO #2 and mineral amendments to all the buckets. I have successfully made IMO #3 with as little 4 tablespoons of IMO #2.

Step 4. Add water-based mineral amendments in the proportions recommended in the recipes in this book.

Step 5. Mix these mineral amendments thoroughly into the water by stirring the water clockwise and then counterclockwise seven times. For a more detailed explanation of proper stirring and mixing, see "Good Water" on page 74. This is a great area of experimentation. Be sure to record what types and how much of each amendment was added at this step.

Step 6. Slowly add the water to the wheat bran. Mixing by hand with friends works well at first while the bran is light and easy to move around. As the contents become wetter and heavier, it is easier to mix the pile using shovels.

Step 7. Continue mixing the water mixture into the wheat bran. The total amount of water required will vary somewhat depending of the initial moisture content of the bran. As the wheat bran absorbs water, it will change color and become darker.

Step 1

Step 5

Step 6

Step 7

Step 10

Step 12

Step 13

Step 8. Check the moisture level periodically. When the moisture level of the bran pile reaches 60 to 70 percent, the mixing process is complete. Use a moisture meter to check the level. If you don't have a meter, tightly squeeze a handful of the bran between both hands. If the bran ball holds together or a small amount of water is squeezed out, do not add any more. Experience is the best teacher here. Remember that the process is forgiving.

Step 9. Next, spread out the pile so that the maximum height is 12 to 14 inches (30–35 cm).

Step 10. Place some small branches atop the pile to ensure good air circulation and then cover the pile, using the bran wrapper bags, leaves, or straw; this helps to conserve moisture within the pile. Try to strategically rip open the bags of bran to maximize their surface area. Place some larger branches atop the cover to prevent wind from blowing things away.

Step 11. Insert a thermometer into the pile close to the center to monitor the temperature, which should rise within a day or two. Monitoring and controlling the temperature in the pile are important. Too high a temperature will destroy important proteins and enzymes. Too low a temperature will cause the pile to become anaerobic. The temperature within the pile should remain between 105 and 120°F (40–50°C).

Step 12. Continue to monitor the pile temperature. If heavy rain is forecast, keep a tarp handy and cover the pile if needed to prevent it from getting too wet. If the temperature rises much above 120°F, use shovels to turn it over to release heat. Adding water is another option to lower the temperature even more if need be. If the temperature falls bellow 105°F, that's a sign that too much water was added to the pile initially. The amount of turning and watering a pile will need depends on many variables, including time of year and amount of sun and rain. Remember to take notes for future reference.

Step 13. After several days white fuzz will appear on the pile. This is the local biology growing throughout the pile and digesting the wheat bran. After about a week the IMO #3 is complete and ready to use in making IMO #4, the final stage of the IMO process.

Uses

I use IMO #3 to make IMO #4. Animals seem to like to feed on IMO #3. I often find a pile disturbed, with parts missing, eaten by local animals during the fermentation process. This product may be stored, but I quickly use it to make IMO #4.

RECIPE AT A GLANCE

▸ Put 150 pounds organic wheat bran on bare soil in shade.

▸ Prepare about 35 gallons good water; dilute fermented local biology (IMO #2) in water 1:1000, and (recommended) add water-based trace mineral amendments.

▸ Mix water into wheat bran pile. Moisture level is correct when squeezed bran ball holds together. If water emerges when bran is squeezed, don't add any more of it to bran (60–70 percent moisture).

▸ Spread out pile to 12 to 14 inches tall, and cover.

▸ Monitor internal temperature, turning pile if too hot (over 120°F).

▸ After about 1 week white fuzz (propagated local biology) will appear; use this to make IMO #4.

IMO #4: Living Soil Amendment

Amendment type: A biological and mineral amendment
Raw materials: IMO #3, soil, rock dusts, homemade mineral
amendments, good water
Content: Diverse local bacteria, fungi, and archaea; other
compounds, minerals

IMO #4 is a living soil amendment that contains a broad spectrum of macro- and microminerals as well as the most diverse local biology available, making it a full complement to any soil, an amendment with profound effects. It is an extremely powerful product that has many uses in the garden and on the farm. Mixing requires a good bit of work, so having several people to share the task, laugh, and tell stories makes the time and effort more enjoyable. There is also solace in the creation of this product. Invoking your intent (see "Mixing Amendments into Water" on page 76) and taking the time needed to complete the task provide for a lovely meditative moment.

IMO #3, with local biology established and digesting throughout the pile, is now able to process larger, more difficult-to-convert minerals, rock dusts, silts, crushed shells, and soil. It is recommended to take advantage of the digestive capabilities of this process. Use amendment math as described in "Garden Math" on page 104 to calculate appropriate amounts of minerals to add. Or just experiment and see what happens. Microminerals may have been added (in the form of water-based amendments) during the manufacture of IMO #3 to address specific needs. These may be added again during this phase as well. The macrominerals contained in rock dusts, shells, and soil may now be added and digested to achieve the mineral proportions desired in the end product. Any of the free or low-cost materials included in this book could be added at this stage.

The first step is mixing the IMO #3 pile with an equal amount by volume of soil. It could be soil from the yard, perhaps the remnants of a recently dug foundation, the garden or soil imported from an outside, trusted source. Be sure the soil added is not contaminated with anything not wanted in the garden. Knowing the mineral constituents of this soil will help you determine what mineral additions are needed. This is true whether your goal is to create a balanced soil mineral amendment or to create an amendment to address specific mineral deficiencies. Conducting a soil test of the added soil before making the batch of IMO #4 and then again after IMO #4 is completed gives insights into what minerals are present in the amendments added. I have done this many times in order to understand how much of an amendment must be added to change the mineral proportions of the initial soil pile. Realizing that the amount of any minerals may be adjusted during this process is empowering.

Adding materials in small amounts at intervals throughout the mixing process results in better distribution of them. IMO #4 is a cornerstone of sustainable, regenerative, local gardening and farming practices.

Materials

IMO #3
Soil
Water-based mineral
    amendments as needed
Rock dusts, silts, shells, other
    amendments as needed
Approximately 25 gallons
    (100 L) good water

Several 5-gallon buckets
Wheat bran wrapper bags,
    leaves, or straw
Plastic cover
Shovel
Thermometer

Instructions

Step 1. Gather all materials needed.

Step 2. Place a pile of soil next to the IMO #3 pile (as made in the previous recipe). The two piles should be about equal in volume; estimating by eye is sufficient. I strain the added soil through ⅜-to-½-inch mesh to remove any rocks. Rock dusts may also be added to the piles.

Step 3. Fill several 5-gallon buckets with about 4 gallons (15 L) of good water each.

Step 4. Mix in water-based minerals, in the manner described in "Good Water" on page 74 in order to get good mixing and energy building.

Step 5. Rock dusts, crushed shells, and other solid materials may be added to the pile during the mixing process. It is best to add these materials in several small batches rather than pouring on a large quantity at one time. I generally add less than 1 gallon (4 L) of rock dusts, mixing dusts from several sources to address macromineral deficiencies of calcium, sulfur, silicon, and manganese. Paramagnetic basalt rock dust is also an ingredient. (See "Rock Dusts" on page 116.)

Step 6. Use shovels to mix the two piles and the water together. This is a slow process of adding rock dusts, mixing the two piles together, adding water, mixing the piles, and repeating. One method of mixing would be to walk around the pile turning shovelfuls over on the edge of the pile and moving the contents into the middle. After a few times around the pile, change the pattern and direction to take shovelfuls from the middle of the pile to the edges, making a hole in the middle. Continue walking around the pile turning shovelfuls over, moving the material back into the middle. Eventually, the two piles become one big pile. This process works whether alone or with others helping out. Mix until a handful of the soil when squeezed just holds together. If a moisture meter is available, add liquids until the moisture content is about 60 to 70 percent. If a small amount of water drips when the soil is squeezed, there is more than enough moisture.

Step 7. Once everything is mixed, spread the pile so that it is about 12 to 14 inches (30–35 cm) tall.

Step 8. Cover with bran bags (as when making IMO #3), straw, or leaves. Sticks may be used to allow air to move freely under the cover. Use rocks or branches to hold the covering in place.

Step 9. Monitor the temperature as described for IMO #3 on page 171, and take steps if needed to cool the pile.

Step 10. Within several days there will be evidence of the white fuzz throughout the pile, and the living soil amendment will be ready for use after about a week.

Uses

The IMO #4 pile should be covered to maintain moisture, and may go dormant at the worst. Store in a bucket inside during the winter for use in early spring before that existing batch thaws or can be made. Experimentation is the best teacher here. Make a couple of batches or more per year. Having seasonal batches of IMO #2 in the refrigerator to make IMO #3 allows you to make IMO #4 with a seasonal variety of biological content.

Remember that when IMO #4 is applied, this mineral- and biology-rich amendment product is alive: It requires moisture, food (organic carbon), housing, and (sometimes) air to survive and propagate when used. Mulch, compost, and cover crops can meet some of these needs! Keeping this in mind helps to harness its full potential.

Distribute IMO #4 systematically throughout the landscape. Here are some starting points to inspire uses for this remarkable product. Remember to record how it was applied, in what quantity, and what results were observed. Unleash your imagination!

- Apply IMO #4 at the end of the day when the application area is out of the bright sun. Consider watering after application if conditions are extremely dry. Consider applying just after a rain.
- Apply a thin layer of IMO #4 on top of the soil a couple of weeks before planting or transplanting or atop mulches to facilitate decomposition.
- Spread thin layers on the compost pile as it is built throughout the season to add needed minerals and inoculate its contents.
- Apply on top of any soil that has a crust that sheds water such as around mulched trees or shrubs. This will open up the soil, allowing the absorption of water. Be ready to mulch again soon, as the biology will quickly digest whatever is there.
- I like to have a pile of IMO #4 on hand for use as needed, especially in the early spring when the biology in the soil needs stimulating. I stash a bucket in the garage to ensure that an unfrozen source will be available in early spring and in the winter.
- Use it to manage the anaerobic floor environment of a chicken coop or pigpen. According to Cho Ju-Young, the microbes in IMO #4 will digest feces and urine, eliminating smells and quickly transforming these wastes into more desirable materials. (It does.)
- I sprinkle IMO #4 like salt or pepper into the soil, mixing it by rubbing handfuls together, at transplanting time or when preparing

a bed for seeds. This requires experimentation, because even small amounts are heat generating.

- To make a powerful biological and mineral tea, put some IMO #4 in a muslin bag, soak the "tea bag" in good water, and then apply the liquid as a foliar spray or drench.

RECIPE AT A GLANCE

▸ Mix IMO #3 with local soil 1:1 by volume.

▸ Mix minerals, liquid, rock dusts, shells, or silts as needed and about 25 gallons good water into pile.

▸ Spread out pile to about 12 to 14 inches tall; cover with paper or leaves.

▸ Cover with tarp in case of heavy rain.

▸ Monitor temperature (110–120°F), turn pile, and add water if too hot.

▸ After 1 week, this living soil amendment is ready for use.

Summary of Amendment Recipes

Table A.1. Raw Materials and Recipes Overview

Name	Type of Amendment	Shelf-Stable	Raw Materials Needed	Content
Water Extractions	Biological then mineral	No then yes	Rainwater, plants	Broad-spectrum minerals and compounds, biology from plant surfaces
Apple Cider Vinegar	Weak acid	Yes	Apples, water	Weak acid for dissolving minerals
Vinegar Extractions	Mineral	Yes	Shells or bones that would otherwise end up in garbage	Broad-spectrum minerals and other compounds
Fermented Plant Juice	Mineral	Yes	Organic brown sugar, plants	Broad-spectrum minerals and other compounds
Fermented Fish	Mineral	Yes	Fish waste, organic brown sugar	Broad-spectrum minerals and other compounds
Leaf Mold Fermentation	Biological and mineral	Yes	Rainwater, leaf mold, plants	Broad-spectrum minerals, biology, and other compounds
Leaf Mold Biology	Biological then mineral	No then yes	A potato (grow your own), leaf mold, rainwater	Local biological diversity
Lactic Acid Bacteria	Biological	Yes (refrigerated)	Water used to rinse rice, raw milk	Broad-spectrum lactic acid bacteria

179

Appendix A

Name	Type of Amendment	Shelf-Stable	Raw Materials Needed	Content
IMO #1	Biological	No	Rice and box	Local biological diversity
IMO #2	Biological	Yes (refrigerated)	IMO #1, organic brown sugar	Shelf-stable local biological diversity
IMO #3	Biological and mineral	Goes dormant	IMO #2, organic wheat bran, homemade mineral amendments	Biological diversity, other compounds, and mineral balancing
IMO #4	Biological and mineral	Goes dormant	IMO #3, homemade mineral amendments, soil, rock dusts	Biological diversity, other compounds, and mineral balancing

Optimal Soil Mineral Amounts

The quantities listed below are "typical" amounts of macronutrients and micronutrients desired in soil. These values are a top-level starting place to strive for as you work on balancing soil minerals. There are ranges for mineral proportions rather than specific values that are not included in this chart, and depending on the type of soil and what is to be grown there, the typical values may be different from what is listed here. Use these values as a reference. Keep in mind that balancing soil minerals may be a long-term task, and that growing high-quality food takes priority over getting the soil mineral proportions right.

For a more comprehensive evaluation of optimal soil mineral levels, see *The Ideal Soil 2014* by Michael Astera.

Table B.1. Optimal Soil Mineral Amounts

Mineral	Target Amount	Form
Calcium	68%	Cation
Magnesium	12%	Cation
Potassium	3%	Cation
Sodium	2%	Cation
Manganese	85 ppm	Cation
Iron	70 ppm	Cation
Zinc	8 ppm	Cation
Copper	4 ppm	Cation
Sulfur	80 ppm	Anion
Phosphorus	100 ppm	Anion

Mineral	Target Amount	Form
Chlorine	40 ppm	Anion
Boron	1.5 ppm	Metalloid
Silicon	75 ppm	Metalloid
Cobalt	1 ppm	Trace
Molybdenum	0.5 ppm	Trace
Selenium	0.5 ppm	Trace
Nickel	0.5 ppm	Trace

A Sample of Dr. James Duke's Phytochemical and Ethnobotanical Database

This table presents the quantities of mineral elements, as measured in parts per million (ppm), of three plants documented in Dr. James Duke's Phytochemical and Ethnobotanical database.

Table C.1. Examples of Nutrient Content of Plants (parts per million)

Element	Dandelion Leaf	Sassafras Leaf	Nettle Leaf
Boron (B)	4.0–125.0	2.0–48.0	6.0–47.0
Calcium (Ca)	1870.0–12,232.0	2,607.0–17,680.0	5,800.0–29,000.0
Chlorine (Cl)	5,300.0–22,000.0	—	1,800–2,700
Cobalt (Co)	—	—	2.6–13.2
Copper (Cu)	4.0–49.0	1.6–102.0	2.0–15.0
Iron (Fe)	31.0–900.0	33.0–1,020.0	8.0–42.0
Magnesium (Mg)	360.0–2500.0	990.0–6,800.0	1,720.0–8,600.0
Manganese (Mn)	14.0–206.0	23.0–1,020.0	2.0–7.8
Molybdenum (Mo)	—	0.0–0.48	0.1–3.0
Nickel (Ni)	0.4–5.6	0.23–3.4	0.3–2.7
Phosphorus (P)	590.0–5268.0	396.0–3,264.0	894.0–4,470.0
Potassium (K)	3,970.0–30,000.0	3,960.0–21,760.0	3,500.0–17,500.0
Selenium (Se)	—	—	0.4–2.2
Silicon (Si)	—	—	2.0–10.3
Sodium (Na)	760.0–5,300.0	—	10.0–49.0
Sulfur (S)	—	—	1,200.0–6,665.0
Zinc (Zn)	21.0–230.0	10.0–136.0	.09–4.7

Source: Data from Dr. Duke's Phytochemical and Ethnobotanical databases, https://phytochem.nal .usda.gov/phytochem/search.

Plant Mineral Deficiency Indicators

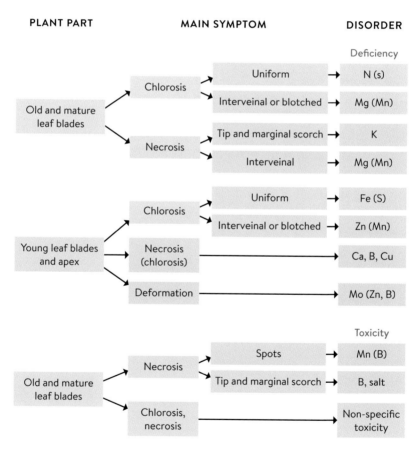

PLANT PART   **MAIN SYMPTOM**   **DISORDER**

Deficiency

Old and mature leaf blades
- Chlorosis
  - Uniform → N (s)
  - Interveinal or blotched → Mg (Mn)
- Necrosis
  - Tip and marginal scorch → K
  - Interveinal → Mg (Mn)

Young leaf blades and apex
- Chlorosis
  - Uniform → Fe (S)
  - Interveinal or blotched → Zn (Mn)
- Necrosis (chlorosis) → Ca, B, Cu
- Deformation → Mo (Zn, B)

Toxicity

Old and mature leaf blades
- Necrosis
  - Spots → Mn (B)
  - Tip and marginal scorch → B, salt
- Chlorosis, necrosis → Non-specific toxicity

Source: Marschner, Petra, ed. *Marschner's Mineral Nutrition of Higher Plants*. London: Academic Press, 2012, p. 300.

Amendment
Mineral Analysis

These tables provide data from the laboratory analysis of several of the mineral amendments I have made. Most of the recipes in this book and many of the plants discussed are represented here. Some plants may be cross-referenced in more than one recipe to give an idea of the mineral extraction capability of different recipes. Successive extractions results are also provided. It is hoped that this data will help those interested realize the wealth of nutrients available at their fingertips from common plants in the garden and backyard, free of charge.

Appendix E

Table E.1. Nutrient Analysis of Water Extraction of Comfrey (parts per million)

	Cl	S	P	Ca	Mg	K	Na	B
	7	0.41	12.71	12.05	5.46	113.5	1.43	0.11

Table E.2. Nutrient Analysis of Vinegar Extractions (parts per million)

Material Used to Make Extract	Cl	S	P	Ca	Mg	K	Na	B
Oyster shells	95	55.04	9.56	6422	84.52	520.4	53.35	2.16
Cow bones	26	24.81	509.2	1691	453.5	618.8	95.44	1.22

Table E.3. Nutrient Analysis of Fermented Fish Extraction and Subsequent Vinegar Extractions (parts per million)

Type of Amendment	Cl	S	P	Ca	Mg	K	Na	B
Fermented fish	1000	127.2	836.8	718.8	105.7	1013	109.3	0.12
First vinegar extract of residue	450	118.6	1137	1177	89.05	825.1	69.09	0.62
Second vinegar extract of residue	175	100.2	1116	1115	55.45	680.1	39.29	1.13

Table E.4. Analysis of Various Types of Leaf Mold Fermentations of Plants (parts per million)

Plant Type Used to Make Amendment	Cl	S	P	Ca	Mg	K	Na	B
Dandelions	33	0.76	5.66	96.47	56.76	197.5	7.99	0.13
Dandelions*	31	1	5.15	84.82	58.25	221.9	8.98	0.13
Grape	7	1.6	39.5	25.15	18.6	309.9	0.53	0.12
Carrot tops	164	22.14	107.9	353.79	68.66	1,055	5.71	1.11

* Values in this row are for the same batch of amendment as the row above. The top-row data represents analysis of amendment taken from the top of the bucket before stirring. This row represents analysis of amendment taken from the bucket after stirring. The similarities demonstrate that there is no need to stir leaf mold fermentations when extracting liquid from them.

Amendment Mineral Analysis

Fe	Mn	Cu	Zn	Al	Co	Mo	Si	Se	Ni
0.51	<.02	<.02	0.07	0.01	0.02	<.01	12	0.09	0.049

Fe	Mn	Cu	Zn	Al	Co	Mo	Si	Se	Ni
0.32	1.32	0.02	0.05	0.26	0.02	0.05	9.61	0.02	0.02
1.39	0.08	0.13	0.14	1.3	0.02	0.02	4.66	0.14	0.02

Fe	Mn	Cu	Zn	Al	Co	Mo	Si	Se	Ni
2.57	1.23	0.17	1.63	1.31	0.02	0.02	0.29	0.16	0.02
3.9	2.25	0.24	2.64	1.75	0.02	0.01	2.28	0.02	0.02
1.31	1.71	0.15	2.34	1.23	<.01	0.02	22.19	0.08	0.02

Fe	Mn	Cu	Zn	Al	Co	Mo	Si	Se	Ni
0.35	0.19	<.02	<.02	0.01	0.03	0.07	6.5	0.13	0.071
0.61	0.26	<.02	0.04	0.13	<.01	<.01	5.9	0.16	0.047
2.39	0.47	<.02	0.12	0.25	0.02	0.02	6.1	0.11	<.02
4.63	1.94	0.04	0.33	0.33	0.02	0.02	3.4	0.13	0.038

Table E.5. Analysis of Various Types of Fermented Plant Juice (FPJ) in Parts per Million

Plant Type Used to Make FPJ	Cl	S	P	Ca	Mg	K	Na	B
Dandelion	1340	33.5	128	143	53.4	485	3.25	0.44
Nettle	1050	70.17	35.34	861	141	376	0.55	1.37
Quack grass root	932	6.32	38.04	98	22	97	1.46	0.12
Sassafras	250	38.7	61.2	245	56	710	2	0.67
Peach fruit	35	5.19	77.06	64.1	21.9	329.7	2.03	0.21
Blueberry fruit	14	6.91	36.4	73.83	21.85	183.3	1.87	0.08
Apple fruit	90	13.01	36.7	95.72	37.02	497.5	6.98	0.32
Horsetail	300	56.79	42.1	358.1	90.92	876.5	1.11	0.22
Lamb's-quarter	65	11.35	153.8	4.5	19.55	903.4	1.29	0.21
Chickweed	250	18.88	205.2	6.84	13.52	1,277	54.47	1.62
Dill	275	46.24	155.2	167.5	55.35	1,157	4.89	0.21
Comfrey	80	8.32	270.8	31.52	34.15	1,025	0.58	0.4
Catnip	125	13.38	43.49	126.6	40.56	331.8	0.78	0.09
Mugwort	200	13.97	62.77	113.4	35.3	16.54	0.88	0.05

Amendment Mineral Analysis

Fe	Mn	Cu	Zn	Al	Co	Mo	Si	Se	Ni
3.17	1.63	0.18	0.56	2.51	0.02	0.13	28	1.8	0.01
1.57	1.18	0.11	0.57	0.9	<.02	0.14	24.6	1.03	0.01
2.54	4.11	0.32	0.28	2.62	0.02	0.03	51.1	1.59	0.01
9.35	3.5	0.39	0.24	3.71	0.01	0.14	30.48	0.37	—
2.77	0.91	0.77	0.17	1.95	0.05	21.9	29.6	0.35	0.02
2.74	1.1	0.7	0.31	1.85	0.02	0.02	25.4	0.26	0.02
6.18	2.77	0.64	0.23	5.37	0.08	0.02	31.3	0.41	0.02
6.74	3.14	0.16	0.65	1.15	<0.01	0.03	28.8	0.3	0.04
5.78	0.42	0.23	0.15	0.38	<0.01	0.03	12.8	0.44	0.021
163.9	1.11	0.31	0.11	3.07	0.02	0.09	13.5	0.21	0.03
6.19	1.14	0.1	0.67	0.45	<0.01	0.06	7.6	0.38	0.04
2.06	0.65	0.15	0.49	0.31	<0.01	0.07	15.4	0.45	0.034
4.56	1.18	0.15	0.4	0.91	0.02	0.05	16.9	0.38	0.03
7.53	1.92	0.19	0.49	1.82	0.08	0.2	21.1	0.54	0.11

Refractive Index Brix Scale

D r. Carey Reams was a pioneer in many fields. This chart was put together by him to establish a method of measuring crop quality and is used around the world. Using a reasonably priced tool, the refractometer, anyone can measure the quality of their fruits and vegetables.

The data in this table represents a range of Brix values defining the quality of some fruits and vegetables.

Table F.1. Refractive Index Brix Scale

Fruit	Poor	Average	Good	Excellent
Apples	6	10	14	18
Avocados	4	6	8	10
Bananas	8	10	12	14
Blueberries	10	14	16	20
Cantaloupe	8	12	14	16
Casaba	8	10	12	14
Cherries	6	8	14	16
Coconut	8	10	12	14
Grapes	8	12	16	20
Grapefruit	6	10	14	18
Honeydew	8	10	12	14
Kumquat	4	6	8	10
Lemons	4	6	8	12
Limes	4	6	10	12

Refractive Index Brix Scale

Fruit	Poor	Average	Good	Excellent
Mangoes	4	6	10	14
Oranges	6	10	16	20
Papayas	6	10	18	22
Peaches	6	10	14	18
Pears	6	10	12	14
Pineapples	12	14	20	22
Raisins	60	70	75	80
Raspberries	6	8	12	14
Strawberries	6	10	14	16
Tomatoes	4	6	8	12
Watermelons	8	12	14	16

Grasses	Poor	Average	Good	Excellent
Alfalfa	4	8	16	22
Grains	6	10	14	18
Sorghum	6	10	22	30

Vegetables	Poor	Average	Good	Excellent
Asparagus	2	4	6	8
Beets	6	8	10	12
Bell peppers	4	6	8	12
Broccoli	6	8	10	12
Cabbage	6	8	10	12
Carrots	4	6	12	18
Cauliflower	4	6	8	10
Celery	4	6	10	12
Corn stalks	4	8	14	20
Corn (young)	6	10	18	—
Cowpeas	4	6	10	12
Cucumber	4	6	8	12
Endive	4	6	8	10
English peas	8	10	12	14
Escarole	4	6	8	10
Field peas	4	6	10	12
Garlic (cured)	28	32	36	40
Green beans	4	6	8	10

Vegetables	Poor	Average	Good	Excellent
Hot peppers	4	6	8	10
Kale	8	10	12	16
Kohlrabi	6	8	10	12
Lettuce	4	6	8	10
Onions	4	6	8	10
Parsley	4	6	8	10
Peanuts	4	6	8	10
Potatoes, Irish	3	5	7	8
Potatoes, red	3	5	7	8
Potatoes, sweet	6	8	10	14
Romaine	4	6	8	10
Rutabagas	4	6	10	12
Squash	6	8	12	14
Sweet corn	6	10	18	24
Turnips	4	6	8	10

Note: All values are degrees Brix, a refractive index of crop sap/juice. 1 Brix = 1% sucrose.

GLOSSARY

Archaea. A newly discovered life-form able to survive in extreme environmental conditions; ubiquitous in soils around the world.

Brix. A measure of plant health calibrated in percent sucrose.

Chelate. A compound in which organic molecules are firmly bonded with metallic ions such as Fe, Zn, Cu, and Mn by means of multiple chemical bonds increasing solubility and supply to plant roots.

Cover crop. A crop of various plant types, such as peas, beans, weeds, oats, and clover, planted between periods of regular crop production to feed the soil biology the nutrients of photosynthesis, prevent soil from drying out or eroding, and after dying back providing humus to increase the exchange capacity of the soil.

Drench. An application of water enriched with nutrients and or biology applied directly to the soil or mulch often in the vicinity of plants for the purpose of increasing the biological diversity and nourishing the root soil ecosystem.

Exchange capacity. A measure of a soil's capacity to hold on to and exchange mineral ions. A high exchange capacity enables availability of mineral ions needed for plant health for the entire growing season.

Fermentation. An anaerobic process of digestion specified by a high concentration of sugar, salt, or other constituent.

Fermented fish. Selected high-mineral fish fermented into a stable form, thereby capturing their mineral content.

Fermented plant juices. Selected high-mineral plants fermented into a stable form, thereby capturing their mineral content.

Float. The fine by-product of a quarry. The "dust" that occurs when rock is crushed so fine that it "floats" on water.

Foliar spray. A fine mist of biological and/or nutrient-rich liquid applied to plants, preferably when they are not transpiring.

Good water. Rainwater or other source of water shown to have a hardness less than 70 parts per million and to be free of unwanted contaminants.

Homeopathy. A system of medicine that treats conditions with minute amounts of natural substances.

Homeostasis. The tendency toward a relatively stable equilibrium between interdependent elements, especially as maintained by physiological processes.

193

Indigenous microorganisms (IMO). The microorganisms (bacteria, fungi, and archaea) living in your backyard or local to your area.

Inoculant. Living material that promotes life to the soil or seed to stimulate the symbiotic relation between roots and soil biology.

Korean Natural Farming. An indigenous regenerative form of farming that utilizes local resources (minerals, biology, and soil structure) to facilitate the local ecosystem producing sustainable agricultural products.

Mother. A gelatinous symbiotic culture of bacteria and yeast that forms on the surface of some fermentations. (Unconditional love!)

Nitrogen fixation. The process of converting the nitrogen in the air (N_2) into forms of nitrogen combinations used in biological processes and ultimately plants and animals.

Plant tincture. A solution of alcohol in which a plant has been soaked to extract minerals and other compounds into a shelf stable form.

Refractive index of plant juices. A calibrated measure (% sucrose or Brix) used to determine plant health. The amount light bends or refracts when passing through a plant juice.

Rhizosphere. The region of soil directly influenced by root secretions and soil microorganisms.

Root exudates. Carbohydrates, organic acids, vitamins, and many other substances essential for life released from plant root systems to feed soil microorganisms.

Secondary metabolites. A range of compounds produced by plants, including flavonoids, terpenoids, nitrogen-containing alkaloids, and sulfur-containing compounds not crucial for plant survival. These compounds are prevalent in healthy plants, and there is interest in their antibacterial, antimicrobial, antioxidant, and anticancer properties.

Stoma. Any of the minute pores primarily in the epidermis of the leaf of a plant, forming a slit of variable width that allows movement of gases in and out of the intercellular spaces.

Tilling. To prepare soil for crops by plowing or digging.

Tilth. A characteristic of the soil that describes the soil's ability to move air, store water, and provide housing for the living organisms in the ecosystem. The physical condition of soil as related to its ease of tillage, fitness of seedbed, and impedance to seedling emergence and root penetration.

Trace elements. Mineral elements that are required by plants for optimum health at very small proportions, parts per million (ppm).

Transpire. To give off water vapor through the plant leaf stomata.

BIBLIOGRAPHY

Anderson, Arden B. *Science in Agriculture.* Texas: Acres USA, 1992.

Astera, Michael. *The Ideal Soil 2014: A Handbook for the New Agriculture, v2.0.* SoilMinerals.com, 2014.

Brady, Nyle C., and Ray R. Weil. *The Nature and Properties of Soils*, 14th ed. Pearson Prentice Hall, 2008.

Brunetti, Jerry. *The Farm as Ecosystem: Tapping Nature's Reservoir—Biology, Geology, Diversity.* Colorado: Acres USA, 2014.

Buhner, Steven Harrod. *The Lost Language of Plants: The Ecological Importance of Plant Medicines to Life on Earth.* Vermont: Chelsea Green, 2002.

Cai, Ran, et al. "The Effects of Magnetic Fields on Water Molecular Hydrogen Bonds." *Journal of Molecular Structure* 938, no. 1–3 (December 16, 2009).

Callahan, Philip S. *Exploring the Spectrum.* Texas: Acres USA, 1994.

Callahan, Philip S. *Paramagnetism: Rediscovering Nature's Secret Force of Growth.* Texas: Acres USA, 1995.

Callahan, Philip S. *Tuning in to Nature: Infrared Radiation and the Insect Communication System.* Texas: Acres USA, 2001.

Cho Ju-Young. *Natural Farming Agriculture Materials.* Translated by Kyung hyum Kim and Hawang In-sung. Seoul: Cho Han-kyu CGNF, 2010.

Cho, Youngsang. *JADAM Organic Farming: The Way to Ultra-Low-Cost Agriculture.* Translated by Rei Yoon. Daejeon, Korea: JADAM, 2016.

Coats, Callum. *Living Energies.* Dublin: Gateway, 2001.

Datnoff, Lawrence E., Wade H. Elmer, and Don M. Huber, eds. *Mineral Nutrition and Plant Disease.* Minnesota: American Phytopathological Society, 2018.

Duke, James. Dr. James Duke's Phytochemical and Ethnobotanical databases. USDA, https://phytochem.nal.usda.gov/phytochem/search.

Fukuoka, Masanobu. *The One-Straw Revolution.* Translated by Larry Korn, Chris Pearce, and Tsune Kurosawa. New York: New York Review of Books, 2009.

Kervran, Louis C. *Biological Transmutations.* Translated by Michel Abehsera. North Carolina: Happiness Press, 2005.

Krasil'nikov, N. A. *Soil Microorganisms and Higher Plants.* Translated by Y. Halperin. Israel Program for Scientific Translations, 1961.

Lovel, Hugh. *Quantum Agriculture Biodynamics and Beyond.* Texas: Acres USA, 1994.

Lowenfels, Jeff. *Teaming with Nutrients: The Organic Gardener's Guide to Optimizing Plant Nutrition.* Portland and London: Timber Press, 2013.

Lowenfels, Jeff, and Wayne Lewis. *Teaming with Microbes: The Organic Gardener's Guide to the Soil Food Web.* Portland and London: Timber Press, 2010.

Mabey, Richard. *Weeds: In Defense of Nature's Most Unloved Plants.* New York: HarperCollins, 2010.

Marschner, Petra, ed. *Marschner's Mineral Nutrition of Higher Plants*, 3rd ed. London: Elsevier, 2012.

McCaman, Jay L. *Weeds and Why They Grow.* Michigan: Jay L. McCaman, Box 22 Sand Lake, 49343, (616) 636-8226, 1994.

Pedersen, Mark. *Nutritional Herbology: A Reference Guide to Herbs.* Indiana: Wendel W. Whitman, 1998.

Phillips, Michael. *Mycorrhizal Planet: How Symbiotic Fungi Work with Roots to Support Plant Health and Build Soil Fertility.* Vermont: Chelsea Green, 2017.

Pollack, Gerald H. *Cells, Gels and the Engines of Life.* Seattle: Ebner & Sons, 2001.

Pollan, Michael. *The Botany of Desire: A Plant's-Eye View of the World.* New York: Random House, 2001.

Pottenger, Francis M. Jr. *Pottenger's Cats: A Study in Nutrition.* California: Price-Pottenger Nutrition Foundation, 2005.

Proctor, Peter. *Grasp the Nettle: Making Biodynamic Farming & Gardening Work.* New Zealand: Random House, 2004.

Schatzker, Mark. *The Dorito Effect: The Surprising New Truth About Food and Flavor.* New York: Simon and Schuster, 2015.

Schauberger, Viktor. *The Energy Evolution: Harnessing Free Energy from Nature.* Translated by Callum Coats. Dublin: Gateway, 2000.

Scott, Timothy Lee. *Invasive Plant Medicine: The Ecological Benefits and Healing Abilities of Invasives.* Vermont: Healing Arts Press, 2010.

Shacklette, Hansford T., and Josephine G. Boerngen. *Element Concentrations in Soils and Other Surficial Materials of the Conterminous United States.* Washington, DC: US Government Printing Office, 1984.

Shepard, Mark. *Restoration Agriculture: Real-World Permaculture for Farmers.* Texas: Acres USA, 2013.

Solomon, Steve. *The Intelligent Gardener: Growing Nutrient-Dense Food.* British Columbia: New Society, 2013.

Thun, Matthias, and Christina Schmidt Rüdt, eds. *The North American Maria Thun Biodynamic Calendar 2014.* Edinburgh: Floris Books, 2014.

Tompkins, Peter, and Christopher Bird. *The Secret Life of Plants: A Fascinating Account of the Physical, Emotional, and Spiritual Relations Between Plants and Man.* New York: Harper and Row, 1973.

Walters, Charles. *Minerals for the Genetic Code.* Texas: Acres USA, 2006.

INDEX

Note: Page numbers in *italics* refer to figures and photographs. Page numbers followed by a "t" refer to tables.

ABOUT THE AUTHOR

Michelle Mullein

Nigel Palmer is a lifelong gardener who relies on the amazing complexity of nature to inspire his gardening philosophy. He is an aerospace engineer by trade, sorting, organizing, and resolving complex technical issues. Nigel is the instructor for the gardening portion of the year-long holistic health education program at The Institute of Sustainable Nutrition (TIOSN), a school founded and directed by his wife, Joan Palmer, an educator, nutritionist, and herbalist. Nigel enjoys telemark skiing up and down mountains, paddling with bears, the keeping of bees, and watching the garlic grow. Find out more at www.Nigel-Palmer.com.

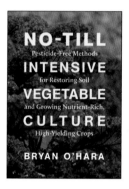
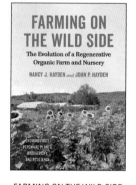
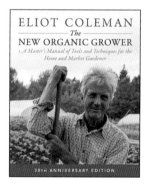
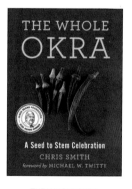